W9-AJO-789

# Women After All

Also by Melvin Konner

*The Evolution of Childhood: Relationships, Emotion, Mind*

*The Jewish Body: An Anatomical History of the Jewish People*

*Unsettled: An Anthropology of the Jews*

*The Tangled Wing: Biological Constraints on the Human Spirit*

*Medicine at the Crossroads*

*Childhood: A Multicultural View*

*Why the Reckless Survive and Other Secrets of Human Nature*

*The Paleolithic Prescription: A Program of Diet and Exercise and a Design for Living* (coauthored with S. Boyd Eaton and Marjorie Shostak)

*Becoming a Doctor: A Journey of Initiation in Medical School*

*Being There: Learning to Live Cross-Culturally* (coedited with Sarah H. Davis)

# Women After All

✦

## Sex, Evolution, and the End of Male Supremacy

Melvin Konner, M.D.

W. W. Norton & Company
New York London

Printed in the United States of America
First Edition



For information about special discounts for bulk purchases, please contact W. W. Norton Special Sales at specialsales@wwnorton.com or 800-233-4830

Manufacturing by Courier Westford
Book design by Iris Weinstein
Production manager: Julia Druskin

ISBN 978-0-393-23996-6

W. W. Norton & Company, Inc., 500 Fifth Avenue, New York, N.Y. 10110
www.wwnorton.com
W. W. Norton & Company Ltd., Castle House, 75/76 Wells Street, London W1T 3QT

1 2 3 4 5 6 7 8 9 0

*In memory of*
*Irven DeVore*
*anthropologist, mentor, friend,*
*and for*
*Susanna, Adam, Sarah, Logan, Becky, and Ethan,*
*who are the future*

"All these arguments we have to-day to offer for woman, and one, in addition, stronger than all besides, the difference in man and woman. Because man and woman are the complement of one another, we need woman's thought in national affairs to make a safe and stable government."

—Elizabeth Cady Stanton, *January 19, 1869*

# Contents

# Women After All

# *Introduction*

✦

## "Stronger Than All Besides"

This is a book with a very simple argument: women are not equal to men; they are superior in many ways, and in most ways that will count in the future. It is not just a matter of culture or upbringing, although both play their roles. It is a matter of biology and of the domains of our thoughts and feelings influenced by biology. It is because of chromosomes, genes, hormones, and nerve circuits. It is not mainly because of what your mother taught you or how experience shaped you. It is mainly because of intrinsic differences in the body and the brain.

Do these differences account for all the ways women and men differ? No. Are all men one way and all women another? Also no. Are the differences I will be talking about affected by experience? Up to a point, yes. But none of these considerations, or many others I will take into account, seriously impedes my argument or deflects its key conclusion: women are superior in most ways that matter now.

And no, I do not mean what was meant by patronizing men who said this in the benighted past—that women are lofty, spiritual creatures who must be left out of the bustle and fray of competitive life, of business, politics, and war, in order to raise the next generation of people with values. I mean something like the opposite of that. I mean that women are fundamentally pragmatic as well as caring, cooperative as well as competitive, skilled in getting their own egos out of the way, deft in managing people without putting them on the defensive, builders rather than destroyers. Above all, I mean that women can carry on the business of a complex world in ways that are more focused, efficient, deliberate, and constructive than men's, because women are not frequently distracted by impulses and moods that, sometimes indirectly, lead to inappropriate sex and unnecessary violence. Women are more reluctant participants in both. And if they do have to be drawn into wars, these will be wars of necessity, not wars of choice, founded on rational considerations, not on a clash of egos escalating out of control.

This is not a new idea. Elizabeth Cady Stanton, who cofounded feminism, gave an address to the National Woman Suffrage Convention in Washington, D.C., on January 19, 1869. She said, "The same arguments made in this country for extending suffrage . . . to white men, native born citizens, without property and education, and to foreigners . . . and the same used by the great Republican party to enfranchise a million black men in the South, all these arguments we have to-day to offer for woman, and one, in addition, stronger than all besides, the difference in man and woman. Because man and woman are the complement of one another, we need woman's thought in national affairs to make a safe and stable government."

She also said, "When the highest offices in the gift of the people are bought and sold in Wall Street, it is a mere chance who will be our rulers. Whither is a nation tending when brains count for less than bullion, and clowns make laws for queens?" Almost 150 years later, the highest offices are still bought and sold on Wall Street, and clowns make laws for queens. But the latter, at least, is coming to an end.

Yet notice: What additional argument for women's equality is "stronger than all besides"? *"The difference in man and woman."* Men and women complement each other. After a century and a half of research, Elizabeth Cady Stanton's argument from difference is stronger than ever, grounded in evolution, brain science, child psychology, and anthropology. And we can take it one step further.

In addition to women's superiority in judgment, their trustworthiness, reliability, fairness, working and playing well with others, relative freedom from distracting sexual impulses, and lower levels of prejudice, bigotry, and violence make them biologically superior. They live longer, have lower mortality at all ages, are more resistant to most categories of disease, and are much less likely to suffer brain disorders that lead to disruptive and even destructive behavior. And, of course, most fundamentally they are capable of producing new life from their own bodies, a stressful and costly burden in biological terms, to which men literally add only the tiniest biological contribution—and one that in the not-too-distant future could probably be done without. As we will see, there are species that have evolved that way and do very nicely, thank you; with our growing mastery of biotechnology, we could get there much faster than they did and do it even better than they do.

I am not recommending that, you understand; I am merely saying it is feasible, whereas the reverse situation is a biological impossibility. As a man, I would like for me and my kind to be able to stick around and maintain some sort of usefulness. I also have it on pretty good authority that most women would not like to get rid of men. But the vast majority of women, and many men as well, want to see a world in which opportunities, responsibilities, and rewards are shared more equally between the two sexes. This is not because they are so similar—although they are in many ways—but, above all, perhaps because they are different. It is the complementarity Elizabeth Cady Stanton spoke of, and I will show that it is not only possible or probable; it is now happening very quickly and is unstoppable.

This is in part because it is inherent in human nature—it represents a return to a distant past, early in our species' history, a complementarity prevented by oppressive social arrangements for thousands of years. That complementarity is returning because it is natural and we are putting those strange social arrangements behind us. Extreme male domination is an anomaly in our history—a long-lasting one but, nevertheless, temporary. In terms of the relationship between the sexes, we are recovering equality, not inventing it. And if we look around the animal world, we find many excellent models among our closest relatives and beyond.

But what will be different about this new world?

Contrary to all received wisdom, women are more logical and less emotional than men. Women do cry more easily, and that, too, is partly biological, although certain male politicians and other prominent men seem able to deploy tears strategically in public. But life on this planet isn't threatened by women's tears; nor does that brimming salty fluid cause poverty, drain public coffers, ruin reputations, impose forced intimacies, slay children, torture helpless people, or reduce cities to rubble. These disasters are literally *man*-made. They result from men's emotions, which are a constant distraction to them. Not all times and places have been the same; men have mastered themselves to various degrees, sometimes for long, even historically long periods. But, as I will show throughout this book, too many are stung by distracting motives and emotions that women are far less affected by.

I have been told that I am too hard on men—that I should recognize that most men are not guilty of violence, rape, promiscuity, or warmongering. Of course they're not. But the minority that is guilty of those things is dangerously large—many times larger than it is in women—and that minority has put a very strong stamp on human history. In fact, you might say it is largely responsible for history, at least for the ten or twelve thousand years preceding modern times. Not even all men in power have been guilty of all those things, but too many have been guilty of one or more of them. This is what

Simone de Beauvoir meant when she said, in her pathbreaking book *The Second Sex,* "The problem of woman has always been a problem of men." As we will see, the behavior of men at the top, or on their way to the top, has been if anything even more oppressive to other men than to women. Men are hugely overrepresented in positions of power, but even more men suffer from those men's behavior; they languish at the bottom, where they are killed, impoverished, starved, cuckolded, and deprived of opportunity by other men.

But—another objection goes—men have accomplished great things! Much more so than women have! I also recognize that, although given that men have blocked women's paths to greatness in all fields for thousands of years, it is scarcely a fair comparison. So let us concede at the outset: most men are not destructive, and not all women are cooperative and nurturing; women have their own means of creating conflict and oppressing others. But in science we ask whether generalizations are possible. I will show you that in the domain of sex differences in brain and behavior, they are not only possible but fully justified by the evidence. We will see, as well, that it is not just a difference in average levels of violence and egotism that has made all the difference in history. It is also that when men get together in groups that exclude women, their higher average levels of these emotions produce a toxic dynamic that has poisoned the stream of history.

In fact, as we will see, all wars are boyish. People point to Margaret Thatcher, Indira Gandhi, and Golda Meir as evidence that women, too, can be warlike. But these women were perched atop all-male hierarchies confronting other hypermasculine political pyramids, and they themselves were masculinized as they fought their way to the top. Also, they were not average women. There is every reason to think that a future national hierarchy staffed and led by women, in a context in which women no longer have to imitate men to lead, dealing with other nations similarly transformed, would be less likely to go to war. But that's not all.

Sex scandals, financial corruption, and violence are all overwhelmingly male. This is not, I will argue, mainly because men happened to be in charge and had the chance to do these things. It is mainly because they are men. And the motives and inclinations that led them into positions where they could abuse power are the same ones that long enabled them to keep women out. But this is over.

Certainly women have emotions of their own: caring for others, especially the young and the old; longing for a future for their children that is better than the past; revulsion at the kinds of things that many men do. But even positive emotions can be overdone in a dangerous world. And women's emotions are hardly all positive. Mean girls have caused other girls to kill themselves just by deploying words. Grown women ostentatiously display wealth to humiliate other women. And women's tactics in the mating game are not all admirable. But even these emotions do not typically threaten the world and its citizens. And the emotions experienced by most women do not narrow, label, and destroy. They broaden, protect, and build, and women know that to do these things they have to live in the here and now, not in some apocalyptic dream. If men are from Mars, women are from Earth.

There is a birth defect that is surprisingly common, due to a change in a key pair of chromosomes. In the normal condition the two look the same, but in this disorder one is shrunken beyond recognition. The result is shortened life span, higher mortality at all ages, an inability to reproduce, premature hair loss, and brain defects variously resulting in attention deficit, hyperactivity, conduct disorder, hypersexuality, and an enormous excess of both outward and self-directed aggression. The main physiological mechanism is androgen poisoning, although there may be others. I call it the X-chromosome deficiency syndrome, and a stunning 49 percent of the human species is affected.

It is also called maleness.

My choice to call being male a syndrome and to consider it less normal than the usual alternative is not (as I will show you) an arbitrary moral judgment. It is based on evolution, physiology, development, and susceptibility to disease. Once in our distant past, all of our ancestors could reproduce from their own bodies; in other words, we were all basically female. When biologists ask why sex evolved, they are not asking rhetorically—the fact that sex feels good was a valuable addition. What they are really asking is: *Why did those self-sufficient females invent males?* It had to be a very big reason, since they were bringing in a whole new cast of characters that took up space and ate their fill, not to mention being quite annoying, *but could not themselves realize the goal of evolution: creating new life.*

We'll consider this in chapter 2, but briefly, the best answer to the puzzle seems to be: to escape being wiped out by germs. When you make new life on your own, you basically clone yourself, and ultimately lots of your offspring and relatives have the same genes. The germ that gets one of you gets you all. Create males, and in due course there is much more variation. Mate with a male that's a bit different from you, and you produce a creature different from both of you. Result: germs confounded. Meanwhile, you export the fiercest part of the competition. You do the reproducing, he doesn't (except for his teensy donation), so he can duke it out with the other males and they can evolve faster. Your daughters inherit the variation, and they compete and evolve, too.

But it turns out you have created a sort of Frankenstein monster, after a certain point hard to control. Consider the lowly, graceful water striders that scoot over pond tops in summer. Females signal that they are ready to mate by causing ripples of a certain frequency to billow out in the water, and the ripples turn males on. But the females don't take all comers. Female choice is vital. Males that don't rate, they drive away. Yet males have their ways. They have evolved grasping antennae, perfectly shaped to get a grip on the female's head. A male approaches from behind and secures his hold, then flips her and himself upside down. Using his rear legs, he positions their bodies. If he gets this far, she stops resisting. He is the one. Or one of the ones, at any rate. She mates several times a day and seems to play males against each other.

This is not an allegory of human mating; it is an illumination, more parallel than parable. Female choice is crucial in humans, too, but males didn't evolve grasping antennae. They evolved strategies of seduction, including romance, patience, persistence, gifts, help, verbal praise, argument, promises, threats, family influence, and deception. Human females have protected themselves with skepticism, social alliances, and a tendency to stay aloof and keep men guessing. The man who talks the best game has usually convinced himself first, and (unlike the water striders, which do it physically) you might say they emotionally flip for each other. Sometimes males use force. In this they rely on superior physical strength, gained through eons of competition with other males for access to those very selective females.

Women, of course, compete as well, against men and among themselves, also with skills honed over eons. But the need to reproduce, with all its risk and cost, has kept them relatively levelheaded and dubious of men's schemes. For most of the history of sexual reproduction, females have often stood by while males fought over them, physically or otherwise. They know that they won't always be able to tell a lifelong pal from a sperm donor, and in many species one good sperm is all they want. But they, too, have to reproduce, and that means tolerating uncertainty and being prepared for contingencies. For us humans, the trouble is that men's competitive antics and untold ages of imposing their will on women have created a world in peril from their rivalries. Females, whether water striders or women, might be forgiven for looking back with a jaded eye on whichever ancestor it was that gave birth to the first male.

Women have always had to struggle for equality, even in the small hunter-gatherer bands we evolved in. Yet with further cultural evolution, it got worse. With the rise of what we like to call civilization, men's superior muscle fostered a vast military, economic, and political conspiracy, enabling them to exclude women from leading roles. Jealousy of women's power to give sex—and, more importantly, to give life—led men to build worlds upon and against them for millennia. Or as Camille Paglia put it in *Sexual Personae,* "Male bonding and patriarchy were the recourse to which man was forced by his terrible sense of woman's power." Appealing myths about Amazons are just that: myths. Only women whose fathers, sons, or husbands gave them the scepters of power could wield it, and then only temporarily. Even in matrilineal societies, men had most of the power. The result was ten millennia in which we squandered half of the best talent in the human race. Brawn mattered for those one hundred centuries, but in spite of their greater strength, men had to make laws to suppress women, because on a truly level playing field, women were destined to compete successfully and very often win.

That is the other meaning of the quote from de Beauvoir: "The problem of woman has always been a problem of men." Although I don't agree with her that all differences between men and women are culturally determined, I fully accept that the majority of the differences we have seen throughout history are caused by male supremacy and the subordination of women. History is written by the victors, and the victors in the battle between the sexes have for many centuries been males; of course they have defined women downward and have invented and promulgated an "essential" inferiority of women as a part of femininity itself. That is the part that is not at all inherent in biology; rather, it is, literally, a man-made myth.

But millennial male dominance is about to come to an end. Glass ceilings are splintering into countless shards of light, and women are climbing male power pyramids in every domain of life. Even in the world's most sexist societies, women and girls form a fundamentally subversive group that, as communications technology shows them other women's freedoms, will undermine age-old male conceit and give them the sway of the majority they are.

The freer and more educated girls and women become, the fewer children they have; men are proven obstacles to family planning. Even in the poorest lands, the increasing availability of women's suffrage, health services, microloans, and savings programs, is giving them control over their destinies. As soon as that happens, they reduce the size and poverty of their families. It becomes clearer every year that the best way to spend an aid dollar in the developing world is to educate and empower women and girls. The consequences are manifold.

Replacing quantity with quality in childbearing will not save just women, or even just struggling, impoverished countries. It will save the planet and make it habitable for our species. It will greatly reduce the necessity for violence of all kinds, as it has already begun to do. Male domination has outlived any purpose it may once have had. Perhaps it played some role in our success as a species so far, but now it is an obstacle. Empowering women is the next step in human evolution, and as the uniquely endowed creatures we are, we can choose to help bring it about.

I've been asked why I, a man, should be writing a book like this. First, I have two daughters and a stepdaughter, all now grown. I love them a lot. I don't like the way the deck has been stacked against them throughout history. I want to make my little contribution to improving the world they must live in. I also have a son. I love him, too. He is not in the minority that exploits women. But I want him, like my daughters, to live in an egalitarian world. I also want them to understand that there are differences between the sexes that are not shaped by culture but are more fundamental, rooted in evolution and biology. I don't want any of the four of them—or my hundreds of students a year, or any young people, or anyone at all—to live with the great disadvantage of missing that fact. As for women, especially young women, especially my daughters, I don't want them to hate and fear men, but I want them to understand that men are often not what they seem. Neither are women always what they seem, but they are less dangerous.

Second, many women have written books explaining what is wrong with men and what men have done to hurt women throughout history. Feminism itself is in large part a declaration of men's wrongs against women. Certainly, it also declares that one of the wrongs is that men have underestimated women—in fact, have consistently lied about women's abilities—throughout that same history. So women are hugely superior to the claims men made about them for all those millennia. Yet a woman writing a book about why women are superior to men can obviously be accused—however unfairly—of special pleading. However good the evidence and argument she lays out, some man will say she's just peddling her own wants and needs. It seems to me that a man who can honestly say the same thing, and prove it, is making a different contribution simply because he is a man. I am a man making an unrepentant argument that women are on average and in aggregate better than men, and I will define what I mean by that in great detail. What can unequivocally be said in men's favor is that they have tried to protect women from other men (for their own purposes mainly, and necessarily failing much of the time) and they have accomplished much more in the realms of exploration, invention, discovery, creativity, and leadership—a fact that obviously cannot be separated from their ancient, chronic, and systematic refusal to let women try these things.

Third, I've been teaching and writing about biological and medical anthropology for over four decades. My work has focused on scientific approaches to human nature and on the evolutionary, biological, and cultural foundations of childhood. I spent two years studying infant and child development and other aspects of life among hunter-gatherers of the Kalahari in Botswana, and I have great respect for the role of culture, but from that and many other experiences I learned that culture cannot explain everything, particularly when it comes to sex differences. However, I also learned that in the most basic human societies women held their own, in many ways much better than they would in the great civilizations that followed.

In the clinical years of medical school, I saw the differences between men and women devolve into illness in different ways, and while men and boys were more subject to most pathologies, women and girls were often in the hospitals and clinics where I was being taught because men put them there—through beatings, rapes, unwanted pregnancies, sexually transmitted infections, and other means. I also helped take care of men who were victims of male violence; in fact, they are the majority of such victims. Most importantly, I delivered thirty-five babies, one of the most inspiring experiences of my life, leaving me with an indelible impression of women's courage and power.

Many of these differences are cultural. But everything I know about the subject tells me that there are durable differences in the behavior and psychology of girls and boys, women and men, that are more fundamental—biological. I will first argue that gender identity is at its core something biological, something set early in life, whether it is masculine, feminine, or one of the many interesting varieties of identity that cannot be simply labeled either. Nevertheless, most people are in some believable sense male or female. Why? To answer this, we have to go to an early point in the history of life to see how and why asexually reproducing cells—basically, females—invented males. We will also have a little tour of the varieties of sexuality in animal species and consider how it was possible for some very complex animals to once again do without males. In the course of this we will begin to understand sexual conflict—that is, rivalry both between and within the sexes—as inevitable and fundamental to life.

We will look at birds, mammals, and, especially, primates—our cousins—in which the conflict ends with females dominating males, others in which the reverse is true, and still others in which the power of the sexes is quite balanced. This will help us understand what happened in human evolution and, in particular, the fact that for most of the hunting-gathering era—that is, most of human history—relations between the sexes were more equal than they were later. This gives us a foundation on which to build the future.

We will then enter into the darker part of history, the thousands of years in which war and preparations for war predominated in human social purposes. This, combined with much larger concentrations of people, enabled men to form coalitions that fully excluded women for the first time and demoted them to a private space away from the public sphere. In this long part of history, men killed each other in large numbers, and when they did they seized the conquered enemy's women, with few limits to the number of women a powerful man could use for his own purposes of sex and reproduction. The later rise of monogamy, which we will try to explain, at first only modestly restrained powerful men from their attempts to control women. But by a little over two centuries ago, women's voices began to be heard again in a substantive way.

At that point we will lay aside the historical thread for a short while, to return to the culture and physiology of sex differences and their development. But this time we will focus more on behavior and less on identity, more on body and less on mind. Our touchstone will be a classic of the mid-twentieth century by the anthropologist Ashley Montagu called *The Natural Superiority of Women,* which became an important text for second-wave feminists. But I will show you that in our new century we can be even more confident of the thesis of that book, not least because of advances in brain science in the past decade. I will also tell you about the current research on how modern men and women differ in experimental situations that test prejudice, bigotry, group identity, and willingness to go to war, and in surveys, observations, and economic analyses of sexual behavior. This research shows that twenty-first-century men think and behave in ways consistent with the long course of male violence and sexual exploitation throughout history, although male proclivities are now channeled mostly in less dangerous paths.

We will then pick up our thread of history in the modern period and trace the tremendous strides women have made and the difference they are making as they gain influence in various walks of life. But we also have to look at the obstacles to that progress, especially in the developing world, where many of the problems with men that women in the developed world have partly overcome remain a huge burden. I will show you how advances are being made on that front and how we can accelerate this trend everywhere, transforming our species and setting the stage for its next evolutionary advance.

Finally, I will explain how the world will be different and better when women have an equal if not a dominant role in running it. This follows logically from all that has gone before; if you are convinced that men and women are fundamentally different (even if you reject the idea that women are superior), you have to believe that a pervasive rise of women even to just an equal status with men will dramatically change how the world works—improve the world, in my view—in many ways. This prediction is independent of the fact that the small numbers of women who have come to the top so far have not yet changed the world very much or even in some cases been very different from men. And we don't have to just imagine how things will be different, because the science already predicts the major outcomes. Although boys and men may have a difficult time with the transition, in the end the new world will be better for everyone.

So of the four main parts of my argument, the idea that important aspects of the difference between men and women in identity and behavior are based in biology will offend many feminists; the idea that these differences as well as the battle between the sexes are deeply rooted in evolution will offend the many Americans who believe (against all scientific evidence) that evolution didn't happen; the idea that women are superior to men will offend a lot of men and some very good people of both sexes who are at present (rightly) worried about boys; and the idea that the success of women is now inevitable will offend those who think (contrary to the facts of history) that conditions for women are as bad as they've ever been and getting worse.

You might say that this book will have something to offend almost everyone. But it will be worth the trouble if it contributes in the least way to women attaining their rightful place after all, and if I convince a few of you that the world will be better when that happens than it has ever been in the past. It is not the end of men, but it is the end of male supremacy, and it is very nearly here.

# *Chapter 1*

✦

# Diverge, Say the Cells

One of the most tragic medical stories of the nineteenth century is that of a French child called Herculine Barbin. It is difficult to write about Barbin because the personal pronoun, so ubiquitous in our language and without which we are tongue-tied—that is to say, he or she, him or her, his or hers—immediately imposes the kind of insistent classification that damaged Barbin's life and finally ended it. We have no language with which to refer to a third sex, even though today we pride ourselves on our sensitivity to people who are gay, lesbian, bisexual, transgender, or transsexual.

Barbin was born in Saint-Jean-d'Angélie, then a town of around six thousand people, on November 8, 1838. The baby emerged with ambiguous genitals, and so the immediate loud cry of "It's a boy!" or "It's a girl!" was probably not heard. No doubt the birth attendants were at a loss for words. Even that "it" with which we precede, for a split second, the lifelong consignment of a newborn to be known as one or the other, to be confined within bars of language, culture, and psychology from cradle to grave; even the word "it"—which would be an insult at any later time—was not an option in French, in which inanimate objects, too, must be either masculine or feminine. We don't know what really happened on that day, but we know something about what happened later. Before committing suicide at thirty, Barbin wrote a clear, affecting memoir condemning physicians and others who, after having left well enough alone for two decades, abruptly forced a choice, ending what was until then a life in satisfactory, or at least workable, gender limbo.

Although in advanced countries the situation is changing, in general we humans have not tolerated ambiguity very well, especially when it comes to male and female. Yet not all—perhaps not most—languages have gendered personal pronouns. For example, in Turkish, *o* means either he or she and *ona* means him or her. It would be interesting to know whether a person of uncertain or intermediate gender would be fundamentally more comfortable in such a language—Persian and Tagalog, the language of the Philippines, are other examples—but clearly these cultures may have definitive, even oppressive sex roles.

Certainly nineteenth-century France had both. Barbin was at first labeled and raised as a girl, so perhaps we can begin by referring to her as such, reflecting her social reality at that time of life. Her middle name was Adélaîde, but her family called her Alexina. She attended a convent school, where she considered herself plain and where she developed a crush on an aristocratic girl. She sometimes slipped into this friend's room at night, and she was punished for that by school authorities. As she entered puberty, she did not develop breasts or begin to menstruate; she grew some facial hair, which she tried to trim away.

Despite these difficulties, Barbin did well enough to go on to a teachers college at age seventeen, and she fell in love with one of the women teaching there. Finally, after joining the faculty at a girls' school, she fell in love yet again, with Sara, a fellow instructor. She liked having Sara dress her, and in time they became lovers. Their liaison came to light, and at around the same time Barbin, always sickly, developed severe pains. The doctor who examined her was not ready for what he saw and suggested that Barbin be fired, which, thanks to the liberality of this particular school, was not done.

At about age twenty-two Barbin went to confession—not new for her, since she was a good Catholic, but it was the first time she revealed how very special she was. Another doctor was asked to examine her. He found a small vagina as well as a small penis, with testicles inside her body. Sometime after that she was deemed legally male; she was forced to leave the teaching job—it was an all-girls school—and to give up Sara as well.

Barbin now became officially known as Abel and briefly made news. He—we perhaps can say that now—moved to Paris and wrote his memoir, but he was unemployed, poor, alone. He eventually sought and found death by inhaling coal gas from the stove in his apartment. Ironically, the address was on the Rue de l'École de Médecine. The memoir, set on the floor near him, was retrieved by medical personnel and lay in the International Office of Public Hygiene for a century, until the philosopher Michel Foucault discovered and published it, together with his commentary, which rightly saw Barbin's story as a tragedy of forced gender identity.

Foucault was a normal gay man who argued, in his three-volume history of sex, that gender is culturally constructed. Many intelligent people today accept this claim. Of course it is true to a large extent, as we know from the considerable cross-cultural variety of sex roles and dramatic historical change we have seen in just the past half century. But if culture alone could explain the caprices of human sexual expression, Barbin would have been able to be a girl and young woman when that was what the culture wanted, then change into a man when the culture's demands changed. Cultural construal—the labels we stamp on people and what we take those labels to mean—can do what it will, but cultural *construction* is something of a misnomer. What cultures do with sex is not designing or building but a finishing process that comes along with and often behind biology. Not long ago, there was a strong belief that to be a nurse is female but to be a doctor is male; that is cultural construction with a vengeance, and cultural change has largely wiped it away. Yet when it comes to gender identity, sexual attraction, romantic feelings, aggression, and perhaps a few other aspects of our lives, cultural construction is more like choosing the cabinets than framing the house. This is a strong claim, and it is part of the purpose of this book to present the case for it.

It is fashionable to say that gender is a continuum and that, therefore, the designation of two sexes, male and female, is arbitrary. If it were arbitrary, then all cultures wouldn't do it the same way, and they don't. Some have a third sex or gender, or even more than three, and for some observers that closes the case. But it doesn't, for the simple reason that all cultures have two very large categories, male and female, and a relatively small category or two for those who don't conform. We can learn from those cultures and face up to the subtlety of gender better than we have so far. If we do, people like Barbin will have happier lives.

But the fact is that they are a small minority.

In one meaning of the word "continuum," we can imagine a long transparent tube, closed at both ends and filled with water, in which two dyes—say, red and yellow—are introduced at the ends. The dyes will diffuse toward each other, and at some point (before the entire tube has become one shade of orange) there will be a uniform continuum in which there is true red at one end, true yellow at the other, and a gradual gradient of oranges in between. If we freeze the tube at this point, we should find equally sized bands of color—about as much middling orange as deep yellow or deep red, which will match the amount of yellow-orange and red-orange, and so on.

But suppose the tube is partly pinched in the middle. If we freeze this pinched tube after the same amount of time, we will find orange in the middle, as before, but there will be much less of it than red or yellow. There will still be a continuum on each side, with red going toward orange as the tube narrows on one side, and yellow going toward orange on the other. This variation on either side of the bottleneck is real and important, but the continuum occurs in a small percentage of the fluid. Yet we could fairly say that the tube has a red side and a yellow side, or a reddish side and a yellowish side, while of the first tube we could only say that it has a red end and a yellow end, with equal amounts of every color in between.

If we claimed that the first tube has a yellow side and a red side, this would impose a dichotomy on a uniform continuum, and a reasonable observer would object. If we tried the same with the pinched tube, making allowance for the shades of red and yellow, and acknowledging that in and around the narrow waist of the tube there is orange fluid that should not be labeled red or yellow—or even reddish or yellowish—we would not be distorting reality, and a reasonable observer would allow us to use, carefully, the two main color words.

Some observers of the spectrum of human gender seem to believe that male and female are part of a myth, because we do indeed find everything in between. But human gender is like the narrow-waisted tube, not the uniform one. If we don't force people to be something they are not, if we protect the rights of the small minority who don't fit the common labels at all, as well as the larger minority who don't fit them very well, we can be fair to everyone without denying the obvious fact that most human beings are either male or female, despite huge variation within each category.

The mainly two-part system doesn't work well for many characteristics. If I tell you how bright, saturated, or pretty the color is, you won't know whether I'm thinking of red or yellow. Similarly, if we play a game in which I ask you to guess whether a person I am thinking of is a man or a woman and allow you to ask a single question before guessing, you would not ask whether the person is musically gifted, or outgoing, or lives in a city. Those questions would waste your chance. And despite the consistent sex difference in height, you would be foolish to make "How tall?" your one question; the overlap between men and women's heights is far too great. However, if you ask, "Does the person you are thinking of have a vagina or a penis?" you would be more than 99 percent likely to guess right after getting the answer.

Of course, what we really want to know is whether certain psychological capacities, such as those for ambition, physical violence, and sexual infidelity, are more like musical ability or more like height. We already know that men and women overlap far too much in these qualities for any of them—or, for that matter, most others—to be decisive or even close to it. No behavioral question would be like the vagina-or-penis question. But the difference could still matter.

It's very instructive to look at people who are really in the middle, and who are that way for biological reasons. Among other things, we soon find out that the middle is not really a linear continuum like the waist of the tube but is more like some exotic glass sculpture—small but beautiful and strange.

Barbin was probably what doctors call a pseudohermaphrodite, which does not imply dissembling. Biologically, true hermaphrodites have both working ovaries and working testes; that means they can make both eggs and sperm. This condition is exceedingly rare. More common, although also rare, is the condition of ambiguous genitalia, which can arise in various circumstances in people who are genetically male or female. For them, the vagina-or-penis question does not have a clear answer, or at least not a consistent one throughout life, even though the ovaries-or-testes question does. Such people have greatly helped the rest of us by allowing research that sheds light on how gender more typically develops. We have to hope they have gotten worthwhile insights into themselves in the bargain, although sometimes, in the past, research has worked against them.

At this point you may be wondering why a book about being female or male is dwelling on those who are not quite either. It's because they are the exceptions who prove the rules. They develop as they do because of departures from one of two rather typical pathways, and the scientific analysis of those accidental detours—together with the deliberate ones in countless animal experiments—teaches us a lot about how the typical pathways work. People like Barbin—all the people who don't fit in the two big boxes—are atypical, not abnormal. They have been called experiments in nature, and in a sense they are, provided we don't objectify them or force them to be something they are not. These are moral obligations that precede and transcend any scientific lessons they may be so generous as to teach us.

But before we delve further into their fascinating, sometimes threatened, and often heroic lives, we must look a bit at how maleness and femaleness, evolved over eons, are re-created in each embryo, infant, and child. What are those two classic pathways, from which some of us depart? The die is not cast in the same way in all sexual animals, but we will meditate on mammals. They cast light like that of a candle on our lives, flickering at times, dim at others, always warm, with some hue of gold. We cannot ethically probe ourselves as we can animals, but studies allow inferences, up to a point, from them to us. Here is the story of sexual development that comes out of both kinds of studies.

We humans have forty-six chromosomes, of which (*usually*) two—X and Y—are sex chromosomes. A woman's eggs each carry one X, and a man's sperm are about equally divided between those with an X and those with a Y. The fertilized egg becomes XX or XY, *usually* synonymous with female and male. Many millions of little sperm spiritedly storm the vastly larger egg, but only one gets in before the egg's rim shuts tight. The winning swimmer loses his tail—although we might also say he's lost his head, which breaks off inside, enveloped by the egg. But, unlike the praying mantis male (of which more later), the sperm doesn't offer its head for nourishment; the egg's huge independent mass takes care of that. The sperm's head delivers genes, particles of prized variation, toward the equally gene-rich nucleus of the egg. When they fuse, the sexual destiny of the fertilized egg is set—it is either male or female. So in losing its head, the sperm gains its reproductive future.

There is more to it than this alone, but the Y chromosome plays the key sex-determining role; however, that doesn't make the Y itself a masterpiece. It's a fraction of the size of the X and looks like a stunted sibling; this is not true in all sexual animals, but it is true in mammals. The Y, petite as it is, ordinarily carries one gene that tips the balance: this gene makes the testis determining factor, or TDF, which triggers the growth of testes from organs that would otherwise become (basically) ovaries. The testes take over from there by brewing specifically male hormones—androgens—that mold the rest of maleness. Another Y gene creates a molecule that suppresses female organs.

Other genes on the X and other chromosomes make their mark on sexual development, but the TDF manufactured by the Y is the chemical key that unlocks the androgens. As we saw at the outset, we can think of maleness as a syndrome, a chromosomal defect shared by 49 percent of humans. It does serious damage. It quashes the body's ability to create new life, causes excess death at all ages, shortens life, increases the risk of diseases ranging from heart attack to autism, and causes physical violence, among other symptoms. Most of this is due to androgen toxicity, mainly testosterone poisoning, although estrogen deprivation and other hormonal glitches play a role. But most of it can be traced back to the Y.

Why "toxicity" and "glitches"? Isn't this just rhetoric? No. The mammalian body plan is basically female. If you have just one X (Turner syndrome), you will not be fertile, but you will otherwise be female, as long as you have no Y. If you have two or more X's but also a Y (Klinefelter syndrome), you will not be completely typical, but you will be basically male. There are rare cases of infertility in women who are found to be XY but are insensitive to androgens due to another gene. And a few men seem to be XX under the microscope but are found to have the key Y genes accidentally attached to one of their X's—something that can happen in a slightly awry cell division. Otherwise it's fair to say: the body plan is female unless the Y flips it into maleness.

With this foundation, we can look at Barbin's modern counterparts, and they will tell us much about the *biological* part of the male-female story. I will try to give them gender-neutral nicknames to avoid either backing them into a gender role or pinning them like scientific specimens.

First, we have people born with ambiguous genitals who were XX embryos partly androgenized in the womb. This can happen either because of a naturally occurring condition or because their mothers were given a drug that mimicked androgens. One natural problem is congenital adrenal hyperplasia, and babies born with it—I'll call them "Ahs"—have a genetically altered enzyme in the adrenal gland, so instead of just making its usual hormones (relating to stress and salt), the gland makes excess androgens. Ahs are born with their genitals somewhat masculinized, as well as with significant medical problems that can be treated but not cured. Some other XX babies with ambiguous genitals were exposed to medications related to progesterone. Unfortunately, progesterone has some of the same effects as androgens, a fact not known when the hormone was given to some women to help them keep their pregnancies. The resulting babies—I'll call them "Andras"—also may have partly masculinized genitals but do not have medical problems.

Ahs and Andras are two of the groups that make the "It's a girl!" or "It's a boy!" decree difficult. But often, like Barbin, they've been raised as girls. Some had their small penises removed sometime after birth so that as children they would have fairly typical-looking female bodies. The Ahs also received hormones, plus further ones at puberty, in an attempt to fine-tune their adrenals and prevent masculine development. As for their genitals, psychologists until recently recommended that Ahs and Andras not be left with ambiguous gender identities. Doctors thought, for surgical as well as psychological reasons, that it would be easier to make them girls.

The prevailing theory back then was called "psychosexual neutrality at birth," a fancy way of saying that children form their gender identities purely through experience—how parents and others treat them, how they are taught to conform, how they identify with and imitate adults, and how just looking at their own bodies confirms their sense of who they are. This, said the theory, is why boys (usually) think of themselves as boys, act as boys, and, in due course, choose girls as their sexual partners. Girls, it was thought, become women psychologically because of a very different, complementary set of experiences. Children and adults who didn't fit—cross-dressing boys, "tomboys," gays, lesbians, and others—supposedly turn out as they do because of family psychology while they were growing up, reinforced by peers and school environments.

Today we know that the process, although surely affected by culture, does not depend on it entirely; in fact, culture is not even close to being the key determinant.

The not-very-numerous Ahs and Andras, along with their families, have often entered patiently into studies of gender identity, and it turns out that by many measures they are more boylike in their behavior than their sisters or other matched girls: they pick boys' games, toys, and clothes more often and like dolls and dresses less. In drawing tests, like boys, they tend more often to draw mobile and mechanical objects with dark or cold colors seen from a bird's-eye view, while unaffected girls tend to draw people, flowers, and butterflies in light and warm colors and depict these living things in a row on the ground. As children, the Ahs and Andras are less likely than other girls to say that they someday want to be mothers.

(I am not saying motherhood is a preferable choice; I am only stating some facts about differences in tendencies.) And, in fact, once they're grown up, more of them say that they do not feel like women (although most do), and more fall in love with and have sex with women. They take this developmental path despite having had hormonal treatments, especially in puberty, that gave them more or less typical women's bodies.

Not everyone interprets these facts in the same way, but for many scientists, they confirm what was known from lab studies: not just the body but the brain, too, is bathed in sex hormones in the womb, and in typical males androgens masculinize the brain. In one classic study, pregnant monkeys carrying female fetuses received testosterone only in the latter part of pregnancy, long after the formation of female anatomy. Even before puberty, with its impressive hormonal makeover, male monkeys one to two years of age are much more aggressive in their play than typical females the same age. But exposure to testosterone before birth gave otherwise normal females an in-between level of this monkeying around, known as rough-and-tumble play.

That's one of thousands of experiments with many different species—rats, mice, hamsters, rabbits, ferrets, dogs, and several kinds of monkeys, a whole menagerie. These experiments show shifts in at least three kinds of behavior: aggression (playful or serious), sexual activity, and responses to infants. There is also an impact on brains; male hormones affect only small areas, but these are in circuits involved in sex, nurturance, and physical aggression. The differences in the brain (mainly in the hypothalamus) and in two behavioral categories—aggression and nurturance—are clear before adult sex hormones come into play. In almost all studies, the early exposure to androgens—the exact timing depends on the species—pushes females in a male direction, while castration or anti-androgen treatment makes males more like females. For these reasons, many psychologists and brain scientists now routinely speak of androgenization (or masculinization) of the brain.

Does this prove beyond a doubt that the Ahs and Andras had their brains and behavior masculinized by exposure to androgens or similar hormones? No. Does it, combined with much other evidence, strongly suggest that this is what happened? Yes, but that's not all.

It's not only XX individuals who can be born with undefined genitals. XYs can be too, or at least with very underdeveloped male genitals, among other anatomical problems. One such condition is called cloacal exstrophy, a complex diversion of developing anatomy. Today, techniques of surgical reconstruction allow almost all XY babies with this condition—suppose we call them "Clokes"—to be assigned as male and have penises with at least some functionality. But this was not the case in the past, and so, as was true of Ahs and Andras, Clokes were often assigned to be female and had surgery and hormone treatments accordingly.

Eventually their gender psychology was explored. In a 2004 study by William Reiner and John Gearhart of sixteen of them, aged between eight and twenty-one, two had been labeled male at birth and remained male by all psychological measures. The other fourteen had been assigned to grow up as female and raised as such. But in the 2004 follow-up, eight of the fourteen expressed a desire to be male, nine had mainly male friends, ten had typically male levels of rough-and-tumble play, and all fourteen preferred typically male toys and games. It's important that six remained content with their female gender identity, but the weight of evidence pointed to a significant role for prenatal androgens.

That same year, Reiner published a longer paper with a larger series of patients and more details. Here is an exchange with a thirteen-year-old who had been told eight months earlier the truth about his condition of birth, diagnosis, and early treatment:

> Q: How long did it take you when your mom and dad actually told you . . . how long did it take you to realize that they were telling you the truth and all of that kind of stuff?

BJ: Right when they told me, because now everything made sense.

BJ took a boy's name, tore the flowers off the wall in his pink room, and had his mom repaint it. He stayed out of school for a week and sometimes felt sad, mainly because he was worried about his friends' reactions. According to Reiner, the parents of these kids were often in turmoil, yet they reported that "at times the children seemed to be almost empowered by their declaration" and the interview with BJ about his transition experience "demonstrates a scenario that is typical of many of these children."

Reiner does not suggest that these children, now boys, have an easy future ahead, but he stresses that surgical correction to female anatomy and rearing as girls was a mistake for many of them. Their parents reported good social functioning. Six had at some point spontaneously declared their status as changed to male; seven others had made the transition decisively after their parents decided to inform them, at various times in their development, about their early condition. The transition was typically easier for the child than for the parents. Reiner (whose follow-up team preserved whatever information or secrecy the family had previously established) concludes movingly, "These children adapt to their lives with severe somatic anomalies, pathophysiologic vulnerabilities, and complex medical and surgical interventions from birth. They do not observe their lives; they live their lives." He also says that the research implies "an important role for prenatal androgen exposure in male-typical development, including male sexual identity. Clinical algorithms and paradigms in these children need to be re-evaluated."

This is equally true of XY babies born without a penis or who lost a penis through a botched circumcision, infection, dog bite or other accident, or even abuse. Again, the idea that gender identity and gender-related behavior was overwhelmingly due to social rearing conditions—combined with the fact that it was easier surgically to create something like a vagina than a penis—led to a number of these boys being labeled and raised as girls and getting hormone treatments at puberty to make them as much as possible into women. One such unlucky child, whose identical twin was typically male, was assigned to be female shortly after the accident. A vagina was surgically constructed early on, and female hormones were given in adolescence. It was hoped that these interventions, along with enveloping family and cultural influences, would make this child an infertile but otherwise normal woman.

It didn't. The young woman always felt uncomfortable in her role. Eventually she searched for information about her early life and learned for the first time her own physical history. What she found out came as a great relief to her. It explained a whole life of feeling like a misfit, and she made up her mind to reorient her identity. When she was a girl named Joan, others arranged for her to have surgery and hormone treatments to make her more female. Now *she* decided to become John and as an adult independently arranged to have the reverse surgeries and medical treatments. One of Joan/John's doctors said in an interview later on, "He got himself a van, with a bar in it. He wanted to lasso some ladies." Sadly, two years after his twin brother committed suicide, John was depressed and took his own life. Since his twin had a typical male anatomical history, John's suicide cannot simply be attributed to his unusual one; obviously the causes were complicated.

Some would say this story is a tribute to free will—Joan decides to become John (as many thousands of transsexuals have now done, one way or another) and, with a little help from her surgeons and endocrinologists, crosses a boundary that used to be thought impassable. But in this case, a determining, or at least a very influential factor—male hormones affecting Joan/John's brain before birth—made her uneasy in her assigned role as a girl becoming a woman. The most basic human compassion allows sympathy for her wish to be a man. In another such case—fortunately, they are rare—the penis was lost at the age of two months, sex reassignment and corrective surgery was done by seven months, and the child was raised as a girl. She retained a secure female identity, although she always leaned toward boy-typical toys and games. Interviewed at age twenty-six, she disclosed that she had had an unsuccessful relationship with a man and was in a successful one with a woman; at this time she described herself as a lesbian. So biology has its influence, but it interacts with experience in complicated ways.

What about the growing number of transsexuals—people seeking surgery and hormonal treatment to make them into the sex that they are not? Today, many thousands have had such surgery. Don't they show evidence of pure free will, now for the first time in history able to be exercised in this way? Not really. People cannot legitimately be granted the chance to surgically and medically change their sex unless they can show that they deeply feel, and have pretty much always felt, at odds with their own bodies. That is, they have male bodies but feel that they should be women, or they have female bodies but feel they should be men. Free will? In one sense, yes. Medicine and surgery allow them to choose, doctors and counselors help them, and the laws in enlightened countries don't stop them. But where did their will to change come from? We don't know the biological history of their profound discomfort, their elemental desire to belong to what we call "the opposite sex."

Randi Ettner, a psychologist who has devoted her career to helping such people, has written, "Many young children experiment with trying on clothes of the opposite sex. Few, however, express the persistent wish to *be* the opposite sex. These children go to sleep praying that they will awaken and be miraculously transformed." The response of the environment is typically harsh. "Few things are as devastating to parents as learning that their child is a transsexual. The narcissistic injury in having a child who wants to change genders is beyond description. Parents can accept a child who is a criminal more easily than a child who is transgendered. Our society reinforces this intolerance." Neither the many cases Ettner and others describe nor the transsexuals' own accounts in autobiographies and interviews give credence to the idea that some kind of aberrant child-rearing experience is the key to understanding what makes a person want this.

But what does make sense, as a hypothesis, is that some personal biological history—genetic, neural, hormonal, or pharmacological—made these people wish to be different from what, in their bodies, they had always appeared to be. It's a hypothesis we can test in the future when we can look in enough detail at brain images of people who want to change sex. A 2011 study by Alicia Garcia-Falgueras, Lisette Ligtenberg, and others found that male-to-female transsexuals who came to autopsy and had their brains examined fell between women and men in the cellular and chemical structure of a part of the hypothalamus involved in sex and reproduction. This needs further study, but such evidence may become part of the case people make for changing gender; objective brain differences could support what they so strongly feel. This should never be a requirement—that would merely be more medical arrogance—but it could very well help them and us to understand. We should have, I strongly believe, the freedom to choose what we want, but that doesn't mean we have freely chosen to want it, or that we can wish it away because others want us to fit society's mold.

Finally, consider the wonderfully interesting people who start out as XY but are born with the fairly unambiguous claim "It's a girl." Many cluster in a handful of villages in the highlands of the Dominican Republic. They come from inbred families and share a recessive mutation that changes a very simple enzyme. It's called 5-alpha reductase, and their syndrome is named for that deficiency. There are also cases in New Guinea, Turkey, and elsewhere, but the best studies have been in the DR.

Julianne Imperato-McGinley, an endocrinologist at the Cornell Medical Center in New York, studied eighteen of these remarkable people, publishing the results in the *New England Journal of Medicine* in 1979 and following up for many years after. She found out that they were called *machihembra* (man-woman) or *guevedoce* (testicles-at-twelve), because of what was known about them in retrospect. These children were almost all recognized at birth as girls, assigned to female roles throughout development, and not viewed as special until they *should* have started to go through female puberty. First, they didn't develop breasts; then their clitorises enlarged and became small penises; finally, they developed broad, muscular shoulders instead of laying down body fat over curved, broadening hips.

In other words, these girls became men. But surely more than a decade of being raised as girls could not allow them a successful transition to masculinity? That guess would be wrong. Of the eighteen *machihembras* closely studied, seventeen made an effective psychological transition.

How is this possible?

First, no known cultural-determinist theory of gender development can explain it. Every strongly cultural theory predicts that these people should have confused identities for life. They do have some difficulties making the transition, but almost all of them succeed. They get married. With a little help from a fertility clinic, most of them have children. They become men of their culture, husbands, and fathers, despite the fact that they were not raised to be that at all.

Our best explanation is biological.

The enzyme they have insufficient quantities of, 5-alpha reductase, converts testosterone to another androgen called dihydrotestosterone (DHT). DHT normally causes a penis and testicles to develop in the fetus; since these fetuses don't have it, they emerge with female-appearing genitals at birth. But what causes male *puberty* is mainly testosterone, and they do have normal amounts of that; therefore, despite their very different starting point, they are able to go through something closely resembling male puberty as the usual changes in the brain's regulation of the sex organs take place in the teenage years.

So why are most of them able to make the transition? Our best guess is that their brains have been exposed to typical male levels of testosterone in the womb. As with the Ahs, Andras, and Clokes, and even the cases of loss of the penis and surgical change to female anatomy, testosterone reaching the brain before birth has prepared the *machihembras* to think and feel like men someday. That day comes when puberty transforms them, and they are surprisingly ready.

What about the more usual range of gender identities and behaviors? Gay men and lesbian women live a broad and intriguing spectrum of different lives, not easily (if at all) classified or labeled. Many if not most of us will have same-sex intimacies if we are completely cut off from members of the other sex, as in prison. Most who turn to this situational same-sex intimacy revert to heterosexuality if and when the opportunity returns. But at the other end of the continuum of same-sex relations (and there are all sorts in between), there are people who will never feel an attraction except to a person of the same sex, no matter the situation.

Among those people, there are men who feel and act in a strongly masculine way and men who feel and act feminine; there are women who love women but feel and act feminine and women who love women but feel more masculine. There are men who cross-dress and like to play football, and these men may be gay or straight. Some women get married in a white gown to the woman they love, become sensitive and devoted mothers, and go off to fight in armed combat in a distant land. All these variations are normal. And, as with transsexuals, people who feel in these varied ways and combinations of ways often do not want to be told that their feelings are simply a matter of choice—that if they accepted certain religious beliefs or entered into certain kinds of psychotherapy they would cease to have the feelings they have had all their lives and instead have the feelings somebody else thinks they should have. Of course, people can be coerced or persuaded to be celibate and alone, to suppress their true selves, but that doesn't mean that the misguided people "helping" them have changed their feelings—except for making them miserable. The reasons that they can't change their feelings are in large part biological—incompletely understood for now, but biological nonetheless.

It helps to see what happens to an XY person who looks and acts female throughout life, as psychologist Melissa Hines and her colleagues did in a 2003 study. This is true of people who, despite the Y, lack androgen receptors or can't make androgens. If this absence of effective androgens is complete, they show no signs of maleness in their brains, behavior, sexual orientation, or identity. Except for being infertile, they are women, period, and they often adopt babies because they *do* want to be mothers. Hines and her colleagues' direct comparison of twenty-two women with complete androgen insensitivity syndrome but XY chromosomal types with twenty-two matched ordinary XX women revealed no significant differences in any psychological outcome. These XY women are, again, an exception that proves the rule. If the Y is blocked from doing its thing, if the androgens can't have their effect, there is no maleness at all, not in body and not in mind.

But what about the variety within the typical range, people who have nothing unusual in their chromosomes and no ambiguity in their genitals? Yet they may be men who love other men but also go hunting, boys who dress up like girls but have crushes only on real girls, girls who beat up boys and who want to grow up to be sexy actresses, women and men who have always been sexually attracted to both women and men, men who love flower arranging and women's fashions but are intensely heterosexual, women kickboxers who put on lipstick and bat their eyelashes at a man they meet at a dance, and a thousand other combinations and variations that no one can predict from any external characteristics or physiological measurements.

What causes these variations remains mysterious, but evidence strongly suggests that a lot of the causes lie in as yet unknown genes. Why would I confidently say that if the genes remain hidden?

Although locating a gene and tracing the chemical pathway from it to the physiology—nerves and hormones—is now the gold standard for claiming genetic effects, it is not the only standard. For well over a century, scientists have realized that identical twins (who share almost the same genes) are more similar in many measurable ways than same-sex fraternal twins (who have only as much genetic relatedness as ordinary brothers or ordinary sisters). This is true for height, weight, susceptibility to many specific illnesses, longevity, muscle strength, body fat, nearsightedness, hearing loss in middle age, skin complexion, hair and eye color, nose shape, cheekbone prominence, and many other physical features.

It is also true of most behavioral traits and psychological measures, including IQ, verbal, mathematical, and musical ability, athleticism, extraversion or introversion, nervousness, aggressiveness, susceptibility to many specific mental and emotional illnesses, religiosity, political conservatism or liberalism, and many others. We can argue about this or that method of measurement, we can point out that genes interact with one another and that this complicates the analysis, we can recognize the considerable power of the environment, including the culture, but in the twenty-first century we can no longer ignore the fact that identical twins are more similar than nonidentical twins, even if the identicals were raised in different, separate environments. There is a number for something called "heritability" that is sometimes calculated from the degree of correspondence between the twins of the different types. Arguments rage over the way this calculation can or should be done.

I don't think it matters, because that number isn't what matters.

What matters is how similar the twins are. If one of a pair of twins develops schizophrenia, an identical twin will have a fifty-fifty or greater chance of getting the same disease, while if the twin is not identical, the risk will be much lower. In the case of schizophrenia, the difference remains very great even if the twins have been reared apart. This is true of many illnesses, mental and physical, and it is true of many traits, including height, verbal ability, moodiness, and aggression. We don't need to know the "heritability" number, or even care whether or not it is worth the calculation; we just need to know that identical twins are a lot more similar (even if they grow up in different environments) than same-sex nonidentical twins—that's how we know that the thing we measured has been influenced significantly by genes.

Twin studies are a good way to find this out, but there are other ways. We have been aware for decades that the biological children of people with schizophrenia who are adopted away from their parents in the first month of life are much more likely to become schizophrenic than the biological children of non-schizophrenic people who are adopted at the same early age by parents with schizophrenia. The same can be said of many different illnesses and many human traits within the normal range. This includes, by most definitions, masculinity and femininity, and it certainly includes lifelong sexual preferences, which are in many ways independent of other aspects of gender psychology and behavior.

The point is that genes matter a lot. Experience, learning, culture, and environment also matter a lot. But we all accept that. Take two identical twins and give just one of them piano lessons, and you will end up with a big difference in piano playing. But give them both the same lessons, and they will likely end up more similar in ability than two nonidentical twins who also both get the same lessons. Six or seven thousand languages are spoken in the world, and the differences among them are all due to learning. Women in New York once thought it was unacceptably sexual to show their ankles, and millions of women elsewhere today think the same about their hair, while currently on New York's beaches women wear only the skimpiest of bikinis, and on the Riviera they don't bother with the tops. Almost all modern women sometimes or often wear pants—unthinkable and even punishable not so long ago.

All of these differences and countless more are cultural and historical, contingencies of collective beliefs and experience in particular human settings. Any one of them may change again tomorrow, just like the historical fluctuations in skirt, hair, and beard length—at this writing, big fluffy beards are popular in Major League Baseball, a sport that in my youth was conspicuously clean-shaven. We know all this, and it tends to make us think that genes don't matter, but they do. What has to be specified is where and how they matter. The things just mentioned are among the myriad ways they don't. But if you have an identical twin with schizophrenia, you unfortunately have about one chance in two of getting it, while your sibling adopted at birth, who grew up in the same environment as you and your identical twin, has less than one chance in a hundred. This leaves ample room for environmental influences—you have a 50 percent chance of *not* matching your identical twin in schizophrenia—but it's still a huge difference.

It is now clear beyond the shadow of a doubt that for *some* differences between men and women, genetic influence is much higher than for others. Put another way, men and women differ in their genes, and these differences explain a few—just a few—of the gender differences that many of us have attributed to learning. Culture is powerful, but it is not all-powerful.

Let's try an extension of the game we played before. I'm going to pick a person at random out of an online telephone directory—from any city or town in the world—and download a lot of information about them. Then I'm going to tell you one fact about that person, and you are going to guess whether they are male or female. If you are right, I will pay you fifty dollars. If you are wrong, you will pay me the same. Or you can decide not to guess and move on to the next candidate. It's up to you. Don't forget: I pick the person at random, and only then do I find out the fact in question about the person. No manipulations or tricks.

Suppose we start with something trivial. I tell you the person has a penis. Now, you are smart enough to know that this does not guarantee a right answer, but it comes so close that you would have to be the least betting person on the planet to refuse to guess on this one. You are almost certainly fifty dollars richer.

We pick another name and investigate the new person. I say that this person, growing up, liked to play with dolls and now likes to wear dresses to parties. Well, you're not quite as certain as you were last time, but you've got a new fifty-dollar bill in hand, and the chances are still overwhelmingly in your favor. Take a shot. Now you probably have a hundred dollars.

I pick a third person and tell you that when they have sex, they greatly prefer or insist that it be with a man. This time, you are actually taking a significant risk—say, 5 percent or so. But with a 95 percent chance of guessing right, what will you do?

You are starting to like this game. The fourth random person, I inform you, has attention-deficit/hyperactivity disorder. The fifth has been arrested three times for assault. The sixth, heterosexual, has been physically abused by a romantic partner. The seventh is often depressed and has attempted suicide four times without success. The eighth pays regularly for commercial sex. The ninth has never masturbated to orgasm. The tenth likes games or sports that involve violence.

Now, not a single one of these bets is a sure thing. But the game is going on and on, and you would be an utter fool not to guess on every question I have asked so far, since you are very likely to get them right. The people numbered four through ten are overwhelmingly likely to be male, male, female, female, male, female, and male.

In the long run, with these odds, even though you will not win every time, you are going to be a very happy gamer.

Of course, I could tell you other facts. I could say the person speaks Mandarin, or is shy, or likes watching television, or is overweight, or has a pet, or is married. You could guess if you like, and probably in the long run you'd break even with these sorts of questions, but you might as well pass and wait for a fact that *statistically, to a very large extent, throughout the world, distinguishes women from men.* Such facts can meet that standard only because culture does not easily contravene them; they arise in large part from genetic differences between the sexes and unfold through hormonal influences on the brain, beginning before birth.

Consider the power of culture in its natural domain. If a visitor to a foreign land points to what we call a dog and looks quizzical, a helpful native might say *chien, perro, hund, kalb, kelev, gou, kutta,* or any one of thousands of other words that mean the same thing and are arbitrarily arrived at through cultural tradition. The possibilities are infinite, and if the world went on long enough, the variations would be infinite, too.

But if the same visitor asks a knowledgeable person, "By the way, are the homicides in your country committed mainly by women?" there are only three likely answers: yes, no, and about the same as men. If culture determined the answer and men and women were not fundamentally different in physical violence, you would expect experts in a lot of the countries to say "about the same" and for the remainder of people to be about equally divided between "yes" and "no." But the fact is that you will never, ever, anywhere get an answer to that question other than "no." Of course, it may be elaborated upon, as in: *No, you idiot. What kind of question is that? Men overwhelmingly commit the violent crimes everywhere. This is one of the few rock-solid truths of criminology.* And so on.

In fact, in the game we played before, I could pick a name at random from a list of all the adults in a given country, then pick from a different country each successive time. If we travel the world, visiting every country, and you play the same way, with that same list of questions, you will occasionally give back fifty dollars, but in the end you will be rich.

There is nothing in human behavior that isn't variable. Some women are very violent and kill other women or men, although some of the men they kill attack or threaten them first. Women evolved to defend their young, and some of the violence done by women or any female mammal serves this vital goal. Paradoxically, women do most of the killing of infants and children in home settings, but that is because they are the ones who are with them, and such acts are much more likely when an unrelated male is in the home, involved with the mother. In many cultures where male-on-male violence is the norm, some men want and are assigned female roles. An occasional culture will regularly use women as warriors.

There is also huge variation in total killing. The most violent cultures, as in traditional highland New Guinea, had a thousand times the homicide rate (wartime and peacetime combined) as the least violent, such as England today. We in the United States have around ten times the rate in England. Most importantly (believe it or not), violence has been in a steady historical decline across the world for at least hundreds of years. These differences are of the greatest importance, and they prove again and again the power of culture to channel, suppress, or give free reign to biology. But over this huge range of cross-cultural variation—three orders of magnitude, or a thousand-fold—it remains a robust claim: men do the great majority of killings in every culture.

We will trace the consequences of this simple fact for the complexities of history and politics around the world, but before we do we need to turn to Darwin's theory and the realities it predicts and explains.

*Chapter 2*

✦

# Hidden in Darkness

Whatever your gender psychology, certain core aspects of it do not result from your upbringing. Basic gender identity, romantic and sexual orientation, and the tendency to violence are aspects of being male or female (or some blend or middle state) that are strongly shaped by genes, hormones, and the brain. But where did this biology come from in the long course of evolution? It is not just psychology or even neuroscience we are after but the evolutionary background to sex and gender, maleness and femaleness. This will take us far afield, to creatures and societies of the present and the past that will help us understand where our gender biology and psychology came from. We are one species in a vast array of sexual beings, and while we express a particular version of the common themes of sex and gender, it is not the only one. Yet to understand our own foibles, we need to look at other worlds. In particular, we need to figure out where males came from, why they are the way they are, whether they are really necessary, and how they can be better managed if (as is likely) women decide to keep them around.

Let's say you are a female Komodo dragon. Quite a stretch, I know, but let's just say. You are a giant lizard, roughly the size of a small woman, living in a grassy, sandy spot near an Indonesian forest, where a deer or a water buffalo can wander close enough for you to catch its scent on the wind. You rush to it and bite its leg, injecting highly poisonous venom (and, according to some, perhaps also infecting it with bacteria—we all have to make a living), then patiently wait to feast on the hapless beast when it succumbs in a few days. You live in a world of smells, which you pick up on any passing breeze, tasting the odors with the forked tip of your tongue. A devilish device, perhaps, but you have a very small brain, so you don't get the irony, and you certainly don't feel guilty.

Overheated after hunting and eating your huge meal, you need a nap in the shade, where you stay the rest of the day and the next. On the third or some other day, something deep inside suggests to your simple brain your need to reproduce, and it tells your tongue to be on the smell-out for a different aroma: maleness. If you stay receptive, you may gain a curt courtship, a coarse grappling of limbs and tangling tails, a perfunctory positioning of his loins on yours. He'll pop out one of his penises—he has a backup, for good measure—and deposit some of his genes in you, with results that will be yours alone thereafter. You may resist, but the big oaf, probably heavier than you, may simply climb on top of you and weigh you down until he's done. If your body is ready, it's ready.

But suppose you don't catch the scent of a male, and no male is around to catch yours. Are you doomed to a lifetime of egglessness, just downing deer after deer? It turns out you are not. Under some conditions, you can double the number of chromosomes in some of your own eggs or fertilize them yourself with the chromosomes in a polar body, a little bud on the egg cell that's a sort of kid sister from its last cell division. It's not your species' usual reproductive routine, but it might enable you and a few other females to colonize a new island without the help—or the hulking intrusion—of males. Of course, you don't do any of this consciously, but evolution has prepared you to get the job done.

Now suppose you are reduced in size by a thousand-fold and you are an eight-inch-long whiptail lizard, maybe the desert grassland whiptail or the New Mexico whiptail of the American Southwest. We don't have to specify that you are female, because in your new species, being all you can be means being female, period. In fact, scientists call you *uniparens* because fathers are nonexistent, even as sperm donors. Females have daughters, as do their daughters, and theirs, with every hatchling growing up to be a generator of eggs and, thus, the sole begetter of the whiptail world to come.

But consider the *tiger* whiptail, dwelling in semiarid grasslands in Arizona, not so different in size, feeding habits, or behavior, except for one big drawback: this species has males, and females *need* them to make babies. Instead of having all daughters, a female tiger whiptail has eggs that hatch about half males, which, reproductively, can do no more than inseminate the real reproducers. Since these tiger whiptails are wasting half their substance, half their productivity, and half their food on non-reproducers, how in earth's name are they going to compete with a species like yours that is all female?

Darwin said it best more than 150 years ago, in a daringly titled paper on primroses and "their remarkable Sexual Relations"—see, there *is* a way to be daring with primroses—which he read to a scientific society in 1861: "We do not even know in the least the final cause of sexuality; why new beings should be produced by the union of sexual elements, rather than by a process of parthenogenesis. . . . The whole subject is as yet hidden in darkness."

"Parthenogenesis" is Greek for what you were doing as a New Mexico whiptail—"virgin birth." The Parthenon was the shrine to Athena, goddess of reason and strategy. Athena stayed a virgin, but she did not reproduce on her own, and she passed on no intelligence to help us understand the huge wastefulness of sex. Sex didn't figure in her own birth, since she was born out of the head of Zeus, the father god. But in real parthenogenesis, it's males you don't need; female whiptails can send Zeus and his sons packing.

John Maynard Smith, one of Darwin's greatest heirs, called the problem "the twofold cost of producing males." You and your sisters among the New Mexico whiptails could double your numbers in each generation while the tiger whiptails in Arizona stand pat. Or put it another way: if a female tiger whiptail were born with a hopeful mutation, being able to reproduce without males, she, her daughters, and her granddaughters would soon run away with all of Arizona's lizardly opportunities. But sex has even more costs than that: males and females have to find each other, and sexual selection (as we'll see) can easily yield results that run counter to survival.

Yet sex is everywhere in the world of complex animals—plants too, for that matter—and most female-only species descended from ancestors burdened with males. In fact, it's increasingly likely that early eukaryotes (pronounced *you-carry-oats*), mere one-celled creatures with their genes enclosed in a nucleus, already had something like males.

So this is the paradox: Why did sex become so firmly established, and why is it rare for species to get rid of it, despite the huge advantage of going solo? This latter part of the question could have implications for us. Is the answer to Darwin's century-and-a-half-old question even now hidden in darkness?

To begin with, there is more than one answer. One of them can be found by going back beyond Charles to another Darwin, his grandfather Erasmus. That ancestral Darwin wrote not only a scientific treatise but also an epic poem about evolution; contrary to popular belief, the idea was in play for centuries before Charles puzzled out how it works. Erasmus Darwin, a doctor who turned down a chance to become personal physician to King George III (of American Revolution fame), devoted much of his life to studies foreshadowing those of the grandson who eclipsed him. In his big book on plants—*Phytologia; or, The Philosophy of Agriculture and Gardening*—he wrote that

> from the sexual, or amatorial, generation of plants new varieties, or improvements, are frequently obtained; as many of the young plants from seeds are dissimilar to the parent, and some of them superior to the parent in the qualities we wish to possess.

For these reasons, "sexual reproduction is the chef d'oeuvre, the masterpiece of nature."

Perhaps it would take a man to miss the downside of sex, to call it *the* masterpiece of nature. Erasmus was sexually liberal (and liberally sexual) himself, Charles's father being only one of his fourteen children by two wives and then the governess, after he was widowed. He may have had a fifteenth with someone else's wife. But he was a great proponent of women's rights and an enemy of slavery, and he understood what was evolving across the ocean. He wrote Benjamin Franklin in 1787, calling him "a Philosopher and a Friend" as well as "the greatest Statesman of the present, or perhaps of any century, who spread the happy contagion of Liberty." Given his love for infant America, maybe Erasmus didn't trust himself to take care of King George, who was trying to kill it in its cradle.

But Erasmus Darwin's wider fame came from long poems that played a role like the most successful popular science today, although it's hard for us to grasp the popularity of book-length poems back then. Erasmus's most celebrated was *The Loves of the Plants,* which gave a genteelly pornographic account of plant reproduction. Since gardening was a popular women's art, and botany therefore a popular subject of study for them, *The Loves of the Plants* became a not-so-secret source of guilty pleasure, and it was condemned as such. Dr. Darwin describes, for instance, the sex life of *Lychnis,* an

English field flower known as ragged robin, in which bright purple or pink females compete to be fertilized by males:

> *Each wanton beauty, Tricked in all her grace*
> *Shakes the bright dewdrops from her blushing face;*
> *In gay undress displays her rival charms*
> *And calls her wondering lovers to her arms.*

*The Loves of the Plants* had many, some perhaps furtive, women readers, who could not in that era easily or respectably read anything more explicit than this.

But the point is that he got something right about what his grandson would call the question hidden in darkness: by producing young that are different from the parent, sex provides grist for the mill of evolution. Today we say that sex reshuffles the deck of genes whenever sperm and egg form and meet, and the rate of evolution depends on the available genetic variation. But this only pays back part of the twofold cost of sex. And since we know that change is possible in asexual species, and even sexual species may not change for millions of years—*if it ain't broke, don't fix it,* says selection—there has to be more to sex than just speeding up evolution.

A second answer is called the "tangled bank" hypothesis, named after this famously eloquent passage by grandson Charles:

> It is interesting to contemplate a tangled bank, clothed with many plants of many kinds, with birds singing on the bushes, with various insects flitting about, and with worms crawling through the damp earth, and to reflect that these elaborately constructed forms, so different from each other, and dependent on each other in so complex a manner, have all been produced by laws acting around us. These laws, taken in the largest sense, being Growth with Reproduction; inheritance which is almost implied by reproduction; Variability from the indirect and direct action of the external condi-

> tions of life, and from use and disuse; a Ratio of Increase so high as to lead to a Struggle for Life, and as a consequence to Natural Selection, entailing Divergence of Character and the Extinction of less-improved forms.

This is most of the last paragraph of *The Origin of Species,* one of the greatest science books of all time. It sets forth in one sentence—"These laws . . ."—the long and short of his pivotal theory, and we can see how it was presaged by the grandfather he never knew. And in the two famous sentences that follow, Charles showed something of Erasmus's poetic gift:

> Thus, from the war of nature, from famine and death, the most exalted object which we are capable of conceiving, namely, the production of the higher animals, directly follows. There is grandeur in this view of life, with its several powers, having been originally breathed into a few forms or into one; and that, whilst this planet has gone cycling on according to the fixed law of gravity, from so simple a beginning endless forms most beautiful and most wonderful have been, and are being, evolved.

Thus ends *The Origin,* including a less risqué but clear account of sexual reproduction, Grandpa Darwin's "masterpiece of nature."

But Charles D. makes a separate point about the value of variation, on the "tangled bank" he asks us to envision. Its countless forms, each at least slightly different, depend on one another for the simple gift of life. Being unique, each can exploit the bank's entangled opportunities in a slightly different way. In other words, if your offspring are identical to you and to each other, you will each have the same needs and will compete head-on with one another *and* your identical mother. Variation creates elbow room; each offspring lives by slightly different means. Another way to make elbow room, even in sexual species, is dispersal. When an oak spreads her seed on the wind, her offspring compete less with her and one another. But not every species can cast its seed, so variation is needed.

This idea has recently proved true for a species of sea squirt. It's a translucent yellowish or turquoise vase-shaped thing, about the size of your thumb if you go down to the heel of your hand, although the little vase seems to split into two spouts at the top. When a bunch of sea squirts fix their bases close together, as they often do, they resemble a hand-sized blob of gelatinous wavy tentacles. Often found on rocks, boat hulls, and other submerged surfaces, the sea squirt is a pest, which means it's a great evolutionary success, and we now know one reason is that when a female squirt mates with multiple males, she increases the genetic variation in her young. This, in turn, decreases competition among them when, as often happens, they hang around together.

But there is a third explanation of sex, now the most widely accepted: the Red Queen hypothesis.

The Red Queen comes from neither of the Darwins but (indirectly) from a younger contemporary of the grandson, another Charles, named Dodgson—the mathematician better known as Lewis Carroll. "It takes all the running you can do," the Red Queen warns Alice in *Through the Looking Glass,* "to keep in the same place." So the Red Queen hypothesis simply says, "It takes all the evolving you can do to keep in the same place," because your environment keeps changing and, to paraphrase Heraclitus, you can't step onto the same tangled bank twice.

Evolve as you will, your predators are evolving, too. You get faster, they get faster—you'd better get even faster right away. You get tougher skin, they get sharper teeth. You get toxins, they get fancier guts. You get camouflage, they get color vision. You get bigger, they unhinge their ominous jaws. To respond, you need gobs of variation. You need DNA to spare, genes on hold that you aren't using yet but may need for a rainy day. And all this running-to-stay-in-place has not even hit critical. It matters most when we switch from the predators that gobble you up honestly from the outside to the ones that nibble at you insidiously from within.

These are the micropredators—parasites, viruses, germs. And here is what they want from you: no change whatsoever. It takes them long enough to adapt to one individual. Once they've done it, *Basta!* They don't want any more challenges. Suppose Mike (microorganism) X evolves inside one of your daughters and she fights it off with immune cells until she's 99 percent free of Mike's progeny. But they have a population doubling time of, what, fifteen minutes? They are back in nothing flat with a new, resistant strain. So she kills off 90 percent of *them*. But then . . . well, you get the idea. It's why you're taking a different antibiotic now than you did five years ago.

The thing is, if you've reproduced without sex—just cloned yourself—Mike X only has to solve this puzzle once. Ultimately immunity depends on genes. And if you have dozens or thousands of genetically copied sisters, he or she or it has already evolved around what any of them can do. Likewise your mother, daughters, nieces, granddaughters—all are vulnerable. The Red Queen says, *Check*. And then: *Checkmate*.

But suppose you export some copies of your precious genes into a class of offspring that will never themselves reproduce. Keep doing that for enough generations for mutations to make a difference. Now take one of that nonreproducing type—a.k.a. males—and put some of his somewhat different genes together with some of yours. Mix vigorously and simmer. You have found a recipe to make Mike X miserable, a way to resist the Red Queen move after move after move.

The only real drawback is, now you have to put up with Mike Y, who's not a microorganism at all. He's a nonreproductive member of your species who carries a pretty pathetic-looking Y chromosome. He's not exactly a predator, although he does have something in common with predators. He won't eat you—his destiny is in your hands, or at least your inner organs. Under certain circumstances, *you* may eat *him*, but first you need to collect his wayward genes, so unless you have other options, you won't feast on him until after you're almost finished mating. He's a member of your own species, sort of. But his chromosomes differ from yours, enough to accord you a new sort of protection.

So, all along you've thought that germs come from sex?

Not so. In evolution, sex comes from germs.

The black widow spider and the praying mantis have found a way to have their males and eat them, too. They carry the sperm-donor concept much further than even the most feckless Don Juan. Females accept the gift of sperm, then have the rest of the male for dessert. But it's not just the melodrama of sexual cannibalism that draws us toward the mantis's prayerful pose or the widow's web. It's that these species, like the virgin whiptails and Grandpa Darwin's wanton blossoms, are vital scientific anchor points in the web *we* weave to catch the mystery of sex.

The black widow lives in a broad stripe circling the planet, pursuing her venomous trade in all but the coldest climates. She is starkly beautiful, about the size of a raisin sitting on eight long legs, her shiny black shell bearing a distinctive reddish hourglass or studded with two red-orange spots laced with a partial white trim, as if her back were bejeweled with bright, flat mushrooms. The silk she spins is lighter but stronger than steel wire of the same thickness. Her venom, much stronger than a rattlesnake's, can kill a human child; it routinely kills a broad spectrum of insect prey and, by the way, the black widow male. She is solitary and shy the year round, except for this rather harsh mating ritual. She weighs thirty times what he does, so it's not exactly a contest.

But it *is* a courtship. Think of a powerful Amazon queen who has won the heart of a Lilliputian. He steps gingerly onto her web and taps the strands, much like the water striders we encountered earlier, but here it's the males' tapping that sends out a song of love—or lust, at least—in patterned vibrations. If she is suitably moved, she allows him closer. She lets him climb her great body, where he continues to woo with fancy stepping. He places his sperm inside her with a special feeler, which breaks off after he makes his deposit; he'll be a lucky boy if that's the only body part he leaves behind.

Sometimes the hourglass seems to mean his time is running out, but it's just part of her siren song. If it's not his lucky day, she grabs, stings, and wraps him just as she would a mosquito, injects the tidy package with digestive enzymes that turn him into a rich broth, and drinks him down. Aided in part by this delicious, nutritious meal, the widow soon lays a silky, liquid sac bulging with two hundred eggs and weaves a blanket around them, wrapping them neatly and attaching the cocoon to her web. She does this five more times with eggs fertilized by that one brave or crazy male, and in two weeks there are a thousand new baby black widows.

In a 2012 study of a closely related spider, the orb weaver, Klaas Welke and Jutta Schneider found that females prevented from eating their mates had fewer, smaller eggs that survived less well under stress. This is also true of female tarantulas, although if they have already mated once, they may lure a few more males and eat them one by one without even letting them have their sexy moment in the sun. Scientists drily note that tarantula females are "nutrient-limited" and "males are high-quality prey." Another study of orb weavers, called "Safer Sex with Feeding Females," showed that males have evolved ways to try to avoid being eaten, but only some succeed.

Female praying mantises have their own charms and slightly different sexual appetites. Some say the males must be praying that they will survive sex (and they do try to survive), but females, if hungry, just seem to pray for their next male meal. Female mantises are world-class hunters that can kill and eat mice, snakes, and hummingbirds bigger than they are. The female can and often does eat the male after sex, but she doesn't have to wait until he's finished. She can start by biting off his head while he is at it, and (typical male) his decapitated body can go right on with the show. If it sometimes seems that men don't need their brains to have sex, consider the praying mantis male. Talk about a boy losing his head over a girl.

Probably the most complex animal with sexual cannibalism is the octopus. One female lingered outside her den near a Pacific coral reef as a small male mated with her thirteen times in four hours. Octopus males can have an intimate encounter or they can mate at a safe distance, depositing sperm with one of their arms; this male kept his distance throughout. Often he jumped back a few feet, sensing danger. Five times, "she showed only a subtle head bob and a faint darkening of her body pattern. On five occasions, the male blanched white briefly," a sign of fear, yet cautiously came on again.

Thirteen was not his lucky number. Shortly after the last mating attempt—it's not clear whether he actually deposited sperm—she sidled up and forcefully swatted him off a ledge of coral. He released abundant ink in defense. "She quickly pounced and engulfed him in her arms" and swam away with him, taking him into a cavity in the coral, where she finished him off. She spent twenty-four hours slowly eating him. The next morning another small male—discreetly, from behind a rock—extended an arm into her den and mated with her for three hours.

Maybe the unlucky first male never injected sperm in all those tries, and his beloved finally thought, *Since you're not good for anything else, I might as well get a few meals out of you.* But it's also possible that he deposited sperm and she collected more from the next guy, whose destiny may also have been her culinary enjoyment. Of course, she didn't think through these strategies; she didn't have the brain for it, and few if any animals other than us do. But evolution gave her the means to act in her own best reproductive interests, almost as if she had thought it through—and maybe better, since our conscious minds don't always serve us well.

All this gives us a lesson in how the evolution of sex and gender work. The male widow, mantis, or octopus has only one function in life: seed the female. Having done this, he may as well find a function in death—feeding her so she can better nourish the offspring that will carry forward their combined genes. Think of the jobless, depressed human dad who commits suicide thinking that his widow and orphans will collect his pension or life insurance. They don't eat him, but, depending on local laws, they may live for a time on the funds generated by his death. In humans, a male might morosely weigh these grisly gains against his future value as a father. But in creatures with nothing remotely resembling fatherhood, giving your body for the cause may be the best move you can make—especially from your mate's point of view.

Still, a male (regardless of species) has his own interests, which might lead him to want to mate another day, perhaps with another female. Now the accounting changes, and we learn a critical lesson: male and female, joined together to make an offspring, have different interests from each other and, in the end, even from their individual young. This is true not just of creatures with sexual cannibalism but of every sexual species in the great spectrum of life. And this is where females lose control of the males they invented, to the process of divergent or even antagonistic evolution.

Consider the syrupy primordial slime of three billion years ago. The early earth's crust has mostly cooled, and in the process simple molecules have been cooked up into more intricate organic ones. These finally build a string of genetic material and an enzyme or two that can help the string copy itself. By definition, the string prevails by making more like itself, and the inevitable slings and arrows of fortune—cosmic rays, volcanic spray, toxins, and so on—will not break down all of them. If the genetic string—RNA or DNA—can elongate enough to make other large molecules, like proteins, that offer added protection, we will have new complexity and, over hundreds of millions of years, a very simple cell, which will give rise to many different kinds of one-celled creatures.

By this point, living things have begun to deal with the challenges of nutrition, predation, and parasites that might make variety advantageous. It is not yet time for the "masterpiece of nature" to appear full-blown. But some mutants manage to trade genes, and by doing this each increases the variety of her offspring. "Her" is still right, because both partners in this soupy exchange can produce their own copies, out of their own bodies.

Some *non*sexual species have mating types—strains within the species that differ enough genetically to avoid cloning by exchanging genes. This means that although you and your mate (the partner you trade a few genes with) can both produce offspring, you choose her from the type that you are not. Opposites attract, even within a basically all-female species. A likely example is *Trichomonas vaginalis,* known as "trich"—pronounced *trick,* appropriately from the human viewpoint—the protozoan that gives vaginas and penises the wrong kind of itch. Ironically, trich is asexual, except for those gene swaps; yet it may have had a sexual ancestor. The way we know this holds a clue to the origin of sex.

Trich is capable of meiosis (*my-oh-sis*). This bane of high school biology comes down to something simple. Cell division, or mitosis (*my-toe-sis*), yields two identical daughter cells. It goes on in many parts of our bodies all the time. Meiosis differs in that it randomly allots half the genes or chromosomes of the dividing cell to each of the two daughters. This happens when we make eggs or sperm, so that when they join they can create something new. Gene swaps without meiosis involve a risk of mismatch between the parts of the genomes traded and their future genetic hosts. Meiosis elegantly divides each partner's genome in half so that the halves can combine with their complements in the mate; this reduces the mistakes made when mates mix their genes. It's the same principle as egg and sperm, except that the combining halves are not so different. But the advantages of meiosis may help explain why sex evolved and stuck in so many species—it's a better way to mix and match genes and get the gold ring of adaptive variation.

The sexual version of this unfolded two billion years ago among some one-celled organisms, and most plants and animals have stuck with it. But some, like the vaginal trich and the virginal whiptail, somehow became asexual again. It's difficult to know why, but in one rather startling animal a reversal has been followed in real time, and it gives us clues to how sex evolved in the first place.

The creature is the asexual New Zealand mud snail. At least it *used* to be asexual. In a wonderful case of evolutionary change seen in a human lifetime—there are a lot of these, despite what creationists claim—these snails re-evolved sex. They became infected with a worm, called *Microphallus* because it sterilizes the snails—male or female, sexual or asexual. This, needless to say, is bad for the guy or gal and the species.

In the lab, when snails are deliberately infected with this worm, both snails and worms evolve faster—the Red Queen running in place again. But the real power of this research is that it shows that in nature, in New Zealand lakes, a small minority of snails that became sexual outevolved their far more common asexual counterparts. *Microphallus* quickly wiped out the most common asexual clones, and the more resistant ones were outbred by those snails that (re)invented males. So the Red Queen favored the sexual snails, crowding out the offspring of what we might call virgin queens.

Nevertheless, as we've seen, some all-female species do persist in nature. In fact, as Olivia Judson and Benjamin Normark point out in "Ancient Asexual Scandals," some of those may never have been sexual. Other species consist of hermaphrodites—individuals that are both male and female, making both eggs and sperm—and these can teach us a lot about our own battle of the sexes. Ordinary garden snails are an example. In a pinch, they can fertilize themselves ("selfing," for short), but instead they mostly seek an equally versatile partner. Now it gets interesting. Each two-sex snail tries to inject sperm into the other. They mate for hours, so presumably it's fun. But they are competing, even if both succeed.

In one of the strangest mating rituals in nature, they shoot love darts—that's the scientific term—a few millimeters long into each other. The dart looks like a spear point, and when snails mate, one or both will have a dart lodged in its body, like an arrow sticking out of an enemy. The dart usually does no great harm, but it does seem to hurt and can leave lasting damage. So why inject it? Because it's coated with hormone-containing mucus that aids the sperm of the snail that shot it and, at the same time, ups the number of eggs in the mate that takes the hit.

Each snail tries to launch these missiles during sex, having stored one or more darts near the penis, which is attached to each owner's *female* organs. Snail biologist Ronald Chase has said, "Love is coming down to war in a way. Sexual conflict plays out," even between hermaphrodites. Apparently all's fair in the snails' love-war, and it does seem a pretty macho way to have sex. Yet I would argue that we can fairly call both snails female, because each goes on to create new life, laying around eighty eggs in the soil that, with luck, will grow up in a year or two. Also, although it's not ideal, each as a last resort can reproduce alone.

Chase thinks that the myth of Cupid and his love darts came from the ancient Greeks' knowledge of snails. The Greeks were good naturalists and would have noticed this display. Cupid's darts, according to the story, make you fall in love—you are smitten, we often say—and while this can happen to both partners, it is not always equal, and sometimes the one more deeply smitten gets put at a serious disadvantage. This is an asymmetry we have in common with garden snails.

One further case of two-sexes-in-one: the red-tipped flatworm, named for its gorgeous coloring, which includes a red-tipped white stripe down the back of an almost iridescent blue body. It is about two inches long and speeds around coastal bays, reefs, and lagoons in the oceans from Myanmar to Australia. In addition to female reproductive organs, each worm has two penises, which it uses to fence—again, the scientists' word—in the ritual leading up to mating. The two would-be lovers rear up with their back ends on the ocean floor and fence it out in a way that looks hypermasculine, each striking and parrying as best he—she?—can.

The contest—romance?—takes up to an hour; it is described in detail in a paper called "Sex and Violence in Hermaphrodites." The worms are not using weapons that some might claim symbolize penises. They are using their actual penises, two each, slapping them against their partner's—opponent's?—penises and trying to jab at least one of their own into the other's flesh, to stick without getting stuck. They don't have to aim for any special spot or cavity, just pierce the skin. They can inject anywhere, and the stream of sperm will find its way to the other's ovaries, making pale streaks that look like lightning, visible through the worms' translucent bodies.

Thus the war between the sexes—without sexes. Each is at once aggressive and coy, intrusive and choosy. But as Leslie Newman, who coauthored the paper with her colleague Nicolaas Michiels, said, "It is better to stab than to be stabbed." Each dueler is trying to choose by resisting fertilization. This limits mating to only the most skillful rivals and ensures that the victor's offspring will have the same skills. Naturalist William Eberhard called it "selective surrender." If you are pricked, so to speak, you make offspring. On the other hand, if you stab but escape stabbing, you have no flesh wounds to heal and no burden of eggs, yet you pass on your genes.

It is not difficult to see how males might evolve in such a system, and there are other species of flatworms that are conventionally male and female. Those females avoid some males and welcome others, and it's not the most aggressive males who get the prize. Perhaps sex originally evolved from hermaphrodites, though it could also have happened the other way around. The possibility of evolving back and forth seems clear.

Either way, these instances of intense competition during sex itself highlight a key fact: all organisms are to some extent in conflict with all others, no matter how intimate the relationship. Yes, many species have cooperation and even altruism, but those nice behaviors always involve limits. The red-tips are engaged in the most important cooperation, the one at the heart of all evolution itself. Yet each wants the upper hand—or, rather, the upper penis; each has something to gain at the other's expense. And that's in species that are not even split into males and females.

Which brings us back to our old friend the unisexual whiptail lizard. The species didn't go from "male *or* female" to "both"; it just got rid of males. It is highly evolved, with a far more complex brain than that of a snail or a flatworm. In the vast spectrum of life, the all-female whiptails are not much simpler than we are. Yet they unambiguously evolved an end of males. Of around forty-five species of whiptail lizards, in one-third males need not apply. Here we can always say "she," because these species are made up entirely of shes; each is a sorority in which all the sisters make babies on their own.

Almost.

It turns out that these lizards—let's call them parths—have something called "pseudosex." However, if it were happening in people—or, for that matter, in our close ape relatives the bonobos, about which a lot more later—it would be called sex between females, and in these lizards that is obligatory. One female mounts another just as a male would a female in a sexual lizard species. Any female can at some point in time be either mounter or mountee, and each may play either role in varied encounters. There is nothing to insert—and there are no sperm or any sharing of genes.

The mounter, though, is usually either past ovulation or has undeveloped ovaries, while the mountee's ovaries brim with ripe eggs. Once she lays them, *she* can be the mounter in the next same-sex tryst. This transition is made possible by progesterone, which (as in humans) surges in the egg producer, but in parths that same hormone induces mounting. Because the parths are descended from whiptail species that had males, parth females respond to testosterone if you give it to them, even though they don't normally have much. As neurobiologist David Crews, who has studied them for decades, puts it, the whole thing is a "snapshot of evolution."

Without getting too much into the lizard sexual brain, both the mounting females and the males of their two-sex cousins have the same neural activity in the hypothalamus. This is the area at the base of the brain that interacts with the body, including sex hormones. The same locations are stimulated by testosterone and suppressed by estrogen in sexual males as in mounting asexual females; in fact, this happens even in mammals. Not only that, but studies by Brian Dias, in collaboration with Crews, showed basic biological similarities between the brain circuits that handle female-like and male-like sexual behavior. They wrote in 2008,

> Given that the first "sex" was female, and [the estrogen] receptor is the most ancestral sex steroid hormone receptor, it is more appropriate to consider the female and [estrogen] as ancestral, and the male and androgen as derived, states. . . . [This concept] maintains the element of the "male phenotype" being imposed on what otherwise would be a "female phenotype" but extends research in new theoretical directions. If in fact males are the derived sex, it follows that males may be more like females than females are like males. . . . Several lines of evidence . . . support this idea, such as the relative ease of masculinizing animals compared with the difficulty of defeminizing animals and the [experimental] resurrection of males in parthenogenetic whiptails, indicating that the genes of male traits are present in this all-female species.

In other words, although the immediate ancestors of the whiptail parths were sexual, go back far enough and you find not only that the first sex was female but that her original, foundational existence left traces in the chemistry of the brain: estrogen first, androgens as an afterthought, and the elimination of males as—for some lizards, at least—an even better *after*-afterthought. In a sense, the parth females were just using their brains, ancient hormonal adaptations, to recover their original condition.

As in all sorts of males, including human ones, castration in two-sex whiptails greatly reduces or eliminates male sexual behavior. But, oddly, many males of the two-sex form will respond to progesterone as well as testosterone. In fact, progesterone induces "the full suite of sexual behavior" in about a third of castrated males of the ancestral two-sex species. One of the brain chemicals involved in both cases is nitric oxide, the same molecule enhanced by Viagra. In the lizards, the effect is in the brain, not the penis, where Viagra causes erections. All-female whiptails have nothing to erect. Two put their genitals together—the lizard equivalent of rubbing vulvas—and it just takes a few minutes to set the otherwise asexual reproduction of the mountee in motion.

At this point, you may be wondering whether progesterone could help a male perform, which might be a boon to many men, despite perhaps (as with Viagra) being a bane to some women, who might prefer fewer male advances. In addition, if you are inclined to consider all possibilities, you may also be wondering whether, in the long run, we humans could follow an evolutionary path similar to that of the unisexual whiptails. Could *our* future evolution eliminate males, keeping female-only sexual connections? This is not just a science-fiction scenario, for two reasons.

First, we don't have to wait millions of years. We are increasingly in charge of our own genes and will soon be able to guide our evolution. Enabling women to reproduce asexually, with or without intimate female contact, could be on our horizon, and if we are worried about reduced variation—losing the game to the Red Queen or overcrowding the tangled bank—we can always fertilize one woman's eggs with genes from another woman's or find new ways of introducing genes to spice up the variation and avoid the risks of identical cloning.

Second, we might just be accelerating a process that has already begun. The Y chromosome has lost many genes over the last 300 million years, although it has given up just one since we split from monkeys, around 25 million years ago. Also, sperm counts were reported to be plunging worldwide during the 1990s, although this may have stabilized. But these uncertain biological trends aside, sociology and psychology seem to be evolving away from the need for males.

Almost twelve million U.S. families are headed by single parents, of which 85 percent are mothers. It is difficult to tease apart the effects of father absence from those of poverty and other deprivations, but the epidemic of psychopathology that some psychologists predicted in the wake of the absent fathers has not materialized. Fortunately, the declining significance of violence has removed part of the need for a woman to have a man around—although our society needs to do much better in protecting women from their own intimate partners as well as from strangers—and the huge influx of women into the workforce, combined with social safety nets, makes males even more dispensable. This is one of the reasons men in male-supremacist cultures are so queasy about current trends. Many single mothers have unintentionally ended up using men as sperm donors and not much else. And more women are *intentionally* putting males in an even more minimal role, through artificial insemination. Here female choice can come down to leafing through a book of stats on the IQ, height, weight, health, education, athletic history, and behavioral records of donor males—photos, but no strings, attached.

Traditional male activities like boxing and hunting are in decline, and the absence of the draft and growing roles for women in the military have badly dented men's age-old claim to a special status as defenders. Machines have been replacing male muscle for centuries.

More girls than boys are entering college, and more women than men are finishing. If Olympian women can't outrun, outswim, outplay, and outfight their male counterparts, they can easily beat *most* men at all those things and more. How long will it be before women can, for all intents and purposes, do everything men can do? How long a path is it, really, from artificial insemination with donated sperm to doing the same with synthetic sperm, preselected to have X chromosomes only, with future generations "fathered" by sperm carrying genes from another egg?

The legal, ethical, and social obstacles are greater than the technical ones, which will be overcome in decades, not centuries. But before we decide to go down that uncertain path, we should note that female-only whiptails and hermaphroditic garden snails, with their Cupid's arrows, are not the only animal models we might emulate in a brave new world of guided evolution.

Most coral reef fish, although not simultaneously both male and female, can switch from one sex to the other during one reproductive career. This has been documented in many different species, including the gorgeous damselfishes and parrot fishes common in home aquariums. If a male is removed from an otherwise all-female group, hormonal processes kick in to turn the dominant female into a new male. Humans, in a future world, could perhaps stay all female, designating one of them to become male only when collectively wanted or needed.

Or consider the anglerfish, in which females are ten times as long as males and a thousand times heavier. The teensy male seeks a female out, sinks his teeth into her flank, and proceeds to fuse his flesh with hers. Then he withers away until he is little more than a sperm factory, ready to fertilize her eggs when she lays them. Seahorses have a neat arrangement where the female and male pair off, swim in synchrony for a few days, even tangle tails a bit, all of which is believed to coordinate their breeding states. They need to time it so that when the female lays her hundreds or thousands of eggs, the male will be ready for her to squirt them into his brooding pouch, after which he will secrete sperm to fertilize them. Now *he* is pregnant, and she goes off to make some more eggs. It seems to be a roughly egalitarian relationship: *Okay, you get pregnant, I'll drift off, ovulate awhile, and bring back some more eggs.* The male uses prolactin, the milk-making hormone in mammals, to help gestate the eggs, and when they are ready they are born live from his belly by the hundreds, a cloud of minute seahorses billowing in the deep.

There are a number of birds and mammals in which females are larger than males. In some cases, males serve and service females, take care of the young, and respect and fear female dominance and aggression. By mating with a harem of males, each female has offspring even more varied than she would have with one male. Women could achieve this kind of female-dominant reversal more easily than the other options we've considered, without getting rid of males.

But suppose future women don't want a reversal, only real equality or, perhaps, a modest advantage over males, just to keep them in line—without major biological change? With that goal in mind, we will look at one of our ape cousins, the vulva-bumping bonobos, who, while keeping males around for *some* of their sexual dalliances, thus doubling their fun, also keep those same males well in line with the clout of female coalitions. This prevents all the hypermasculine menaces that befall their and our other cousins, the chimps. Yet before women try to choose from this bestiary banquet of future possibilities, they have to decide what males are good for.

*Chapter 3*

✦

# Picky Females, Easy Males

Charles Darwin, well into *The Origin of Species,* dropped "a few words about what I call sexual selection" that echo, often harshly, to this day:

> This depends, not on a struggle for existence, but on a struggle between the males for possession of the females; the result is not death to the unsuccessful competitor, but few or no offspring. Sexual selection is, therefore, less rigorous than natural selection. Generally, the most vigorous males . . . will leave most progeny. But in many cases, victory will depend not on general vigor, but on having special weapons, confined to the male sex. A hornless stag or spurless cock would have a poor chance of leaving offspring.

Darwin clearly understood the consequences:

> Sexual selection by always allowing the victor to breed might surely give indomitable courage, length to the spur, and strength to the wing to strike in the spurred leg, as well as the brutal cock-fighter,

> who knows well that he can improve his breed by careful selection of the best cocks. How low in the scale of nature this law of battle descends, I know not; male alligators have been described as fighting, bellowing, and whirling round, like Indians in a war-dance, for the possession of the females; male salmons have been seen fighting all day long; male stag-beetles often bear wounds from the huge mandibles of other males. The war is, perhaps, severest between the males of polygamous animals, and these seem oftenest provided with special weapons. The males of carnivorous animals are already well armed; though to them and to others, special means of defense may be given through means of sexual selection, as the mane to the lion, the shoulder-pad to the boar, and the hooked jaw to the male salmon; for the shield may be as important for victory, as the sword or spear.

Now, there is plenty here to take issue with. Males struggle for *possession* of the females? The "best cocks" are the ones that are best at violence? Alligators are "like Indians in a war-dance"? None of this would pass in a work of science today. But Darwin offers an upside:

> Amongst birds, the contest is often of a more peaceful character. . . . There is the severest rivalry between the males of many species to attract, by singing, the females. . . . Birds of paradise, and some others, congregate; and successive males display their gorgeous plumage and perform strange antics before the females, which standing by as spectators, at last choose the most attractive partner. . . . I cannot here enter on the details necessary to support this view; but if man can in a short time give elegant carriage and beauty to his bantams . . . I can see no good reason to doubt that female birds, by selecting, during thousands of generations, the most melodious or beautiful males, according to their standard of beauty, might produce a marked effect.

So we have either the gladiator spectacle of males ripping each other to shreds, with the winner taking possession of passive females, or hopeful auditioning males preening and strutting their best stuff, their destiny controlled by picky females. Either way, it's a peculiar theater in which to play out the history of life—"strange antics," indeed.

But tactless wording aside, was Darwin right? Well, there is one thing he was wrong about: sexual selection is *more* rigorous, not less. Of course, death by being snapped up by a predator, succumbing to a parasite, tumbling out of a tree, or starving can't be pleasant. But for the arithmetic of evolutionary destiny, sexual selection is more severe than any of those fates, because you may have reproduced quite nicely, thank you, before you were gobbled up. Yet you can live long and fall short of the next generation's gene pool, in which case (evolutionarily speaking) you might as well not have been born. Evolution is a struggle for reproduction, not existence, and the sole goal of survival is to get to reproduce (or to help close relatives do the same). Life is the handmaid, procreation the queen.

Consider a population of a thousand in which there is no differential survival. Everyone lives exactly sixty years and drops dead on that birthday. No evolution, right?

Wrong. Of the thousand, a hundred have ten offspring, a hundred have none, and the others have every gradation in between. To the extent that any feature even modestly genetic contributed to these differences, that trait and its underlying genes will spread. There doesn't have to be variation in survival, although in real life there is, and it figures in evolution. But you don't need it to evolve. What you need is variation in breeding success, *partly* determined by traits that are *partly* determined by genes. Ironically, the very process of domestication—artificial selection—which Darwin used as a model for his natural selection, does not work through differential survival at all. Farmers and trainers make fatter cattle, woollier sheep, faster horses, tamer dogs, and eggier hens just by deciding which ones to breed.

Darwin went on to write a book about all this, expanding the "few words" of *The Origin*. There, he described the process throughout the animal world, and he proved its importance. But it would take another century, until 1972, before more neutral language was used to set out a general theory. This was in an article called "Parental Investment and Sexual Selection," by Robert Trivers, who put the case this way:

> Where one sex invests considerably more than the other, members of the latter will compete among themselves to mate with members of the former. Where investment is equal, sexual selection should operate similarly on the two sexes. The pattern of relative parental investment in species today seems strongly influenced by the early evolutionary differentiation into mobile sex cells fertilizing immobile ones, and sexual selection acts to mold the pattern of relative parental investment.

In fairness, this summarized a long paper in which the language was not so balanced, but in this passage at least we have a non-gendered account. We start with the idea that one sex invests more in offspring and then reason that *that* sex is a scarce resource for which the second sex will have to compete. The latter will evolve either adaptations for victory in battle—size, muscles, claws, simple horns, and teeth—or things like songs, tail feathers, fancy antlers, manes, dances, or the ability to create bowers or generate light shows, mild electric shocks, or aromas that the first, more valued sex finds attractive.

If you are thinking that this more neutral language is just a way to cover up the same old male chauvinist claims, consider the cassowary, a gorgeous flightless Australian bird that is man-sized and dangerous, with knifelike claws and a kick that has maimed and killed people. That is, the *female* is man-sized; the male is about two-thirds as large. Both sexes have casques (the name derives from the French word for "helmet"), hornlike pointed crests rising up from the big bird's brow like the bronze crest on a gladiator's headgear. The bird's head is pale blue, the neck a lustrous darker blue, grading into a glimmering orange nape from which red wattles dangle. The female is more spectacular, and she attracts her diminutive males with a series of low-pitched grunts. When one shows up, she may display dramatically by splashing in a puddle, raising fountains of spray. They mate for up to two months, and the female lays her eggs—perhaps not all of them his—and moves on in search of other pliable males. He will brood the eggs and chicks all by himself, and for nine months, under his watchful eye, they will learn to make a living in the forest.

Writing in the September 2013 *National Geographic,* biologist and writer Olivia Judson describes a male named Dad:

> His chicks, which are about four weeks old and almost knee-high, make funny whistling-peeping sounds as they run about. He mostly stays silent—but from time to time clacks his bill, making a large banging noise. He burps too. And occasionally he booms. That is, he tucks his head down low, inflates his neck, and makes a series of low booming noises. As he does this, his feathers puff up. When he sits down, the chicks cuddle up to him, often snuggling into his feathers. . . . The chick also picks ticks off its father's neck and eats them. Yum.

Meanwhile the errant mom is off to greener sexual pastures. As one (human) mother of five put it to Judson, "I'm coming back as a female cassowary."

Cassowaries are the biggest birds with devoted dads and hit-and-run moms but not the only ones. Jacanas, also known as Jesus birds, trot on lily pads and so seem to walk on water, but that's not their main claim to fame. What they are most known for is that, as in the cassowary, the sex that invests more in the young is the male. Before the female lays her eggs on her chosen lily pad, she collars a male who is bound to take care of them, and that's the end of her interest. The male? Seduced and abandoned, while the female, twice his size, goes off to mate with another male—as many as four in an hour, thirty in a season. Wham, bam, thank you . . . *sir*?

Only there's not even that much of a thank-you in this water-lily world, where the females are too busy fighting off rivals and mating here, there, and everywhere to bother with the niceties. Their adversaries? Other females, who are trying to steal their males. By the end of the season they've a harem of cute little dads, each slavishly brooding a clutch of eggs, then doting for most of a year over a clump of chicks. But all of the father's effort won't necessarily stop the odd menacing female prowler, who can (when the boy-harem's Amazon queen has her back turned) trot in, cow a pint-sized pa, peck open his eggs, and dispatch his hard-won babes.

And it isn't just the brooding. He may have risked his life to draw a snake or alligator away from them, instinctively flopping around as if he had a broken wing, so the not-too-bright predator would forget to eat his little ones. But now he sees a strange female—twice his size, true, but probably less dangerous to him than the reptiles were—and does little or nothing to stop her from doing in the chicks. Why?

Probably because of what happens next. She displays her (to him) intensely attractive rump, and soon enough he mates with her, fertilizing *her* eggs. She has won the prize away from the first female. The prize? *The sex that invests more in the young.* Although she has the upper wing, she is just, as ornithologist Stephen Emlen put it, an egg-making machine. The future dad instinctively starts pulling together a new nest on the lily pad, and in a week or so the once-menacing intruder returns, lays some eggs, and goes off to her next conquest. In due course, the pond is dotted with single-father families.

Eight species of jacanas circle the globe in tropical zones, and despite some fossils, we don't understand their origins. It's likely, though, that they evolved from pair-bonding birds, of which there are some *eight thousand* species. They, too, tend to have devoted dads but not single ones, since parents share the burden of raising the young more or less equally. And it is a burden.

Watch the little perching birds in your eaves or garden. If they have a nest of chicks, they fly off nonstop in tandem—they practically need an air traffic controller at the nest—to bring back food. When they alight on the nest's edge, they face a bevy of gaping mouths, often colored and marked in ways that, along with incessant peeping, turn mom and dad into doting slaves. If you're a parent yourself, this may sound familiar. If not, take a moment to thank those who slaved away to keep you in earthworms and away from dangerous cats. Humans, we'll see, are largely pair-bonding too, a fate that carries with it a suite of other adaptations under the logic of sexual selection.

In most pair-bonding birds, the sexes are about the same size, and usually neither has evolved extra weapons or great beauty compared to the partner. Each sex may have to fight off rivals, but not to the extent shown by jacanas, where males invest the precious resource that females have to fight for. In pair bonds, parenting is share and share alike, so once they've found each other, the lovebirds can concentrate on the kids. As with the male jacana, this is a full-time job, but in these species it's a full-time job for two. Emperor penguins lay one egg at a time in their frozen Antarctic landscape, and the look-alike male and female take turns warming the egg and then the chick while the other goes out to swim and hunt in the icy sea, returning faithfully to regurgitate fish for the young.

This egalitarian kind of baby care, so common in birds, has the crucial result that male and female have roughly equal reproductive success. Together they mate, brood the eggs, raise the kids—repeat next season. Sure, there'll be stolen copulations, sneaking around on both sides, and even occasional desertions—folklore aside, no pair-bonding species, not even geese, is perfectly loyal. But why not act the female jacana, trotting around corralling male after male, dumping eggs on them, and withholding child support? For *her* there's a huge advantage: she can field dozens of young in a season, while the male yields far fewer of his own offspring.

Only there's a catch. He can pretty much count on that lower number year in and year out. She has a chance at the breeding jackpot but an even greater chance of being squeezed out by a larger, tougher, even vicious female. Some jacanas can have harems of docile, willing males and leave their eggs in good care all over the pond top, but—barring a rare surplus of males—this has to mean other females have none. In fact, it's this greater, all-or-none *variation* in reproduction that made the females big and rough in the first place. Much more than for their meek little males, it's a zero-sum game. The result for females is more competition, faster evolution, and divergence from the size, shape, color, and behavior of the male.

In a great many other species, including large numbers of fish and countless insects, females are larger than males. We saw how the use and abuse of males stakes out an extreme in the black widow, where the female has her way with males and wraps them up for a midnight snack, and the praying mantis, whose females, in a way, pray for males and then prey on them in mid-copulation, not missing a beat. We also encountered pregnant male seahorses and asexual lizards that have completely gotten over the whole male thing. Compared to them, the jacana way is mild. But in birds and mammals, even that much female dominance is rare. There are those eight thousand species of pair-bonding birds, as well as quite a few mammals, on the egalitarian plan. Yet often in birds and generally in mammals, the jacana arrangement is turned upside down.

A peacock struts his stuff slowly, arcing great turquoise plumes that dwarf his glistening blue body, raising a patch of iridescent gold coins, then sweeping a delicate green mesh up into a lustrous fan dotted by gorgeous, staring green-and-gold eyes, in which the bird stands onstage alone, radiating a gaudy spray with feathers like the sun's rays, only in color. Another turn or two later, he enfolds himself in drapery, collapsing his sumptuous feathers down into a sleek, pied multicolored tail that seems to loll along behind him endlessly. A drab female demurely watching this spectacle wouldn't seem to stand a chance, but she does stand her ground, because he's not the only male trying to wow and woo her. In fact, she has her pick of them, prancing, splaying, waggling, and dragging their stunning quills.

Like a breeder with a monocle at a dog show, *she* gets to choose which one's genes are worthy of posterity. She sizes him up and matches him with her own calm and valued self—thus giving him a gene channel, a shot at the zero-sum. Over the eons, peacocks got grander and gaudier because they caught peahens' eyes and the hens said, *You* or, at least, *Oh, as well him as another.* The rest is evolutionary history, the grandest display of male fashion ever to sidle and glide across an earthen runway, as if he's thinking, *Don't pass me by, girl, it doesn't get better than this*, while she's thinking, *Boy, get over yourself.* She may play hard to get or wait for an even spiffier male. Actually, in time she may choose several mates. But maybe in the end this one's not so bad, and after all, she *is* in the mood.

As for just how good a mood she is in, we know, thanks to beautifully designed experiments by Marion Petrie and her colleagues over many years, that peahens lay more eggs for males with larger trains (possibly a product of their hormonal state) and that their offspring grow and survive better after release into the wild *even when the hens are randomly assigned to different males* and the offspring are reared under matching conditions. The suggestion is that a large train is an honest advertisement of overall quality, then imparted to offspring, but it could also be that the most impressive males (who don't do any of the child care themselves) inspire hens to invest more in the young.

The peahen's preference has turned out to be more complex than it seemed at first, but an elegant 2013 study by Roslyn Dakin and Robert Montgomerie proved decisively that one component of the display matters to her a lot: "Our study shows that the blue-green eyespot color overwhelmingly influences peacock mating success." It's when those spots catch the sun at a certain angle that the hens are won over. The authors, who titled their paper "Eye for an Eyespot," are continuing to examine other aspects of this stunning romantic dance—and for the most part confirming Darwin's original surmise.

We've seen something like this process in jacanas, except in peacocks it's the males who compete for the precious parental capacity owned by females. Peacock males contribute nothing to the care of the young. They put their energy into building beautiful tails and fanning them out in females' view. According to zoologist Amotz Zahavi's "handicap principle," such useless or even detrimental appendages—peacock feathers puzzled Darwin mightily, because they seemed such a lure for predators—actually signal quality, because only a superior male could afford the cost and risk. Many studies have now shown that these add-ons do indeed signal male quality by other standards—measures of health, such as parasite load. In this case, the ornament and the antics that go with it are dubbed "honest advertisements" of excellence.

But according to mathematical biologist Ronald Fisher's classic theory of "runaway selection" (also known as the "sexy sons" hypothesis), it may not really matter, because once the preference for prettier tails gets started (even by chance), females will keep choosing them because their sons will be better off having them, and this cycle will strengthen generation by generation. This should be most true in rich environments, where the cost of frills can be more easily borne and the gain for chosen males great. Whether the adornments evolved by handicap or runaway selection, they work; in most experiments, peahens prefer the cocks with more elaborate tails.

Similar effects have been shown in many species, including ones that to us look much less fine than peacocks. Although coloration also matters, tail length in male barn swallows can be manipulated to influence mating success; under otherwise natural conditions, females mated to males who'd first had their tails lengthened are better at brooding eggs and steal fewer copulations with other males. This is why long tails have persisted despite their possible disadvantage in foraging and survival.

In palmate newts, the male's tail ends in a threadlike filament. In their wild little world of European ponds, females watch courtship displays longer when the beau boasts a longer filament, and in experiments both filament length and display duration predict how many sperm masses a female will accept from a given male. In the tiny Trinidadian guppy, females that choose males with longer tails have more sons, and the sons have long tails, too. In the mandrill, a big monkey much more closely related to us than newts or guppies, males have bright red noses flanked by bulging iridescent pale-blue cheeks; Joanna Setchell showed that the more brightly colored the male, the more females hang with him, groom him, and mate with him, even after controlling for his rank on the dominance ladder. The list goes on and on, and even in our relatively pair-bonding species, male features such as broad shoulders, big pecs and biceps, chest hair, square jaws, and five o'clock shadows have been shown to figure in many women's sexual choices.

So in pair-bonding species, female choice remains important, but it follows a different path. First, sexual selection is more balanced; both male and female may be attracted to certain physical traits. The roseate tern is a sleek foot-long flier seen around the world. Both sexes are white underneath, with gray wings, a black cap, and sometimes an orange beak. Both have swallowlike tails with lengthy outer streamers, and each sex fancies long streamers in the other. In one population in Queensland, Australia, a dearth of males led some females to pair with other females, while those with longer streamers got the hard-to-find, coveted males. But despite the pair-bonding, the similarity of the sexes, and the fair balance of parenting, there is one dramatic asymmetrical behavior: courtship feeding. The female tern sits patiently while the courting male goes a-fishing—but not for a lazy afternoon. The busy boy fetches her gift after wriggling gift of silvery prey; the way to her heart is through her stomach. He is, of course, also proving that he can bring home the anchovies for their future young.

Thousands of species of birds do courtship feeding. Male cardinals bring seeds and put them *in* the female's mouth. Gull suitors bring up a half-digested blend of fish and squid, laying the prize grandly at the lady's feet. Kestrels and owls offer freshly killed mice or lemmings to their perching hoped-for brides, while in other raptors males actually pass the bloody package to the female during flight, dropping the bird or snake as the agile mom-to-be flies under him. Despite the equality of parental effort after the eggs are laid, the female still has to manufacture and lay them, and this makes her a goal worth working for—and feeding her improves her nutritional condition and that of the young. It's not until a species gets on the jacana plan that males actually do much more parenting than females. In other species, males have to prove themselves—as foragers, hunters, providers, fathers, sex objects, or fighters. But even in our postmodern, postfeminist world, men usually pay the restaurant bill on dates, our version of courtship feeding.

These rituals do not work by magic; in birds, at least, they are wired into the brain. In the ringdove, bowing and strutting by the male and cooing by both sexes stimulates the male's and female's gonads, a song-and-dance routine that lasts for days and gets both ready for mating. First the male chases the female and she withdraws, but at a certain point he sits on the nest and gives a softer coo, which she returns, further stirring her own hormones. This goes back and forth until, if she accepts him, she mates with him and they raise the young together.

None of this, of course, is conscious or deliberate, as it might be if human males and females were doing some version of it. It's hormonal; it's instinctive; it's evolved. But, conscious or not, our version evolved, too. Men woo in a variety of ways, but only those that persuaded women have descendants around to tell the tale. True, in many cultures and much of the past, men had to persuade women's families, but as we will see, women's choices mattered then as well.

In fact, it is not far from the truth to think of the whole of evolution—at any rate, evolution since sex was invented—as a long, slow, meticulous breeding experiment done *on* males *by* females. Of course, if a female bird or mammal wants to reproduce, she has to choose some male or other, and unfortunately it may not be a dazzling show but superior size and brute force that conclude the issue (as happens the other way round in jacanas). Still, a female has to be willing and ready, and not every male can bring about that readiness.

Even a female rat, called a doe, has to be lured into lordosis, a dramatic reflex triggered by some males. First the doe has to want the buck. Then she darts toward him and flirtatiously hops away, or just skirts him in a move known as a run-by. She'll do this, wait a bit nearby, seductively wiggle her ears, and repeat the sequence. She might even intercept his moves on another doe with her own moves on him. If this seduction succeeds, he'll begin to mount her, and the touch of his legs on her flank will trigger the last enabling reflex. She'll stretch her lower back and lift her butt, so he can easily enter where she now wants him to be.

This is estrus, instinctive and cyclical. At a certain point, when the sequence has gone far enough, you can trigger lordosis with your finger on the doe's flank, but that's not the real world. As with the ringdove, brains and hormones are involved. You can make it happen by stimulating part of her hypothalamus. Giving her estrogen puts the circuits, including the nerves in her pelvis, on high alert. And the response can be contravened: in a famous study brilliantly called "Sex with Knockout Models," Emilie Rissman and her colleagues, using mice, showed that females whose gene for the estrogen receptor is knocked out attack males when they would normally be receptive and fail to show lordosis when their flanks are stimulated, even after estrogen injections. These built-in systems are just part of the meaning of instinct. But there is also choice. All right, the doe isn't sitting around with her friends discussing how good-looking a buck is, how much money he makes, or how often he showers. But rat and mouse do have their ways of choosing, and eons of evolution have honed their instincts for acceptance or rejection according to male quality.

Creatures like peacocks are known as tournament species, because they gather on a breeding ground where males show off while females pick and choose. Female choice here is intense. But unlike in jacanas or pair-bonding species, the females aren't probing for possible doting dads; doting is not on the dance card of the chauvinist males available. So the females are after three things: sex appeal for their future sons; such high quality that the male can waste energy and risk death to build ornaments and show them off in dazzling displays; and, significantly, the ability to intimidate and defeat other males.

This is where the tournament comes in. In many species, the males are like jacana females: not so much ornamented as big and tough, domineering and dangerous, to females as well as to other males. Black grouse are a classic example; males congregate in a spot that has little to recommend it except that females know they can find mates there. The males, almost two feet long, dark gray with iridescent blue highlights and dramatic bright red eyebrows, belt out a distinctive mating song and swagger to exhaustion or until a female—smaller and a drab brown—chooses them. These jamborees happen at dawn in the spring and can involve hundreds of birds. But Anni Hämäläinen and her colleagues, in a 2012 study, showed that it is fighting performance in males that the females are mainly after. Similarly, in the Uganda kob, an antelope around three feet high, males are taller and heavier than females and sport horns.

Many individual males defend territories, where they try to maintain harems; others collect on mating grounds, where three to seven males hold the center and monopolize sex with many more females.

You don't, however, need crowded tournaments to have intense male competition. Consider the majestic red deer, studied for decades on the Isle of Rum, off the west coast of Scotland, by zoologist Tim Clutton-Brock and others. Close kin to the American elk, they are equally imposing and very successful, ranging over Europe and parts of Asia and North Africa. Stags can weigh over 500 pounds and hold aloft an intimidating rack of antlers; hinds are bareheaded but can weigh 350 pounds. Mature stags maintain harems and peak in breeding at age eight. During the autumn rut, rivals challenge them. The stags walk in parallel to assess each other's body and antler size. They often bellow, launching a loud hoarse groan. If they fight, the clash of antlers is colossal, and they can hurt each other badly. Victors gain or keep their harems and approach and mount hinds, even beginning thrusting with several hinds before completing the act. A hind may walk away. The rut is exhausting; stags may shed 20 percent of their weight. It's no surprise that males age out of all this within a few years.

But in these stags and for other deer and antelope, are the horns true weapons, in Darwin's original sense, or just costly ornaments advertising quality? Probably both. Certainly antlers and horns impress choosy females, but they also intimidate rival males. If the rival is of an age and size to do more than put on a show, these accessories can inflict serious damage. But they often don't make sense purely as weapons; they are designed not so much to harm as to inspire. So ornamentation clearly matters. Size itself is vital for male defense, but it is also clearly sexy in many species, and in some species females can "see" it in the dark. The croaks we hear in a kind of chorus at night around a pond are belched out by male frogs broadcasting their presence to females. At the same time, they are declaring their size. Although other factors matter too, the more basso the croak, the larger the frog, and the more likely the female is to approach when she hears it.

Nowhere, however, do size and force matter more than in the elephant seal. To be anywhere near them on a beach is to know how puny we humans are. The relatively svelte females are only eleven feet long and a comparatively modest fourteen hundred pounds; males are three feet longer and more than triple that weight. They have huge swellings on their snouts to help them belt out colossal roars, and they charge across the beach and slam into each other at astounding speeds. They sink their teeth into each other's necks and shoulders, bloodying their chests until they are smeared red and pocked with wounds that leave permanent scars. These epic fights take no notice of any thing or creature in the way, including females and even pups that could be their own, which they sometimes crush to death, like trucks flattening squirrels. Cassowary and jacana females may be as large and cruel on their own scales, but the scale of elephant seals is titanic.

Yet every pound of the male elephant seals' bulk and every thrust of their brutality are worth the cost and risk to them, because the potential reward is huge. As Burney Le Boeuf showed many years ago, 4 percent of the males get 85 percent of the sex. Of the other 96 percent, most get none at all—if they survive. But a winning bull can inseminate fifty cows in a season, fathering so many pups that he can afford to squash a couple while defending his benefits. Plan B is to be replaced at the top of his massive harem by a rival and breed no more.

It's easy to see that these kinds of odds and differences in success could produce rapid evolution, at least for a time. Size can't increase forever, because there are physical and physiological constraints. And the shared genes of males and females set a limit on the divergence between the sexes, especially in birds and mammals. But the process of getting there, of diverging until you have pressed against those limits and then staying there, can follow an obstinate evolutionary logic. Cassowaries and jacanas are exceptions that prove the rule; they show decisively that it's not about masculinity and femininity, it's about parental investment and sexual selection. In mammals, however, the sex that invests more in the offspring is usually the female, which means that most species with divergent sexes are the opposite of jacanas.

In a classic paper, "Mammals in Which Females Are Larger Than Males," Katherine Ralls founded a field of research. She began by noting that females are probably larger than males in most species of invertebrates, many fish, and some amphibians and reptiles. Among birds, she reminded readers of jacanas and phalaropes, but she also said that few biologists seem to be aware that the females-bigger phenomenon occurs in mammals, too. She found eighty-four examples.

Among them were hippopotamuses, okapis, several small deer and antelope, mongooses, hyenas, certain seals, whales and dolphins, several mice and shrews, quite a few bats, a couple of flying squirrels, some hares and rabbits, chinchillas, Burchell's zebra, and a number of small marsupials, including the Tasmanian devil. Among our relatives the primates, she mentions South American monkeys, but we now know that females are larger and dominate males in many species of Madagascar lemurs, as we'll see in the next chapter. Ralls also found that in almost all of the eighty-four species, females exceed males in length by less than 10 percent, with an average of less than 5 percent. This is nothing like the size difference favoring females in many insects and fish or even in cassowaries and jacanas. And it's nothing like the excess males show in many mammals, elephant seals being the most extreme.

However, when we consider that men are only around 7 percent taller than women, the female size advantage in all those other species doesn't seem so trivial. The golden hamster, for instance, reverses our own ratio. And in what is often called the largest species that ever lived on earth, the blue whale, cows are bigger than bulls, the record female stretching almost one hundred feet.

Ralls thought, and recent research confirms, that the reasons for larger females are complex. Some female-larger mammals have a lot of fathering help from their male mates, but most don't. Males do defend the young, but this happens in many male-bigger species too. Nor are larger females more aggressive than smaller males; males usually fight more, as in other mammals. In many of these species, the smaller males have typical Darwinian weapons and ornaments specialized for fighting—against other males. Ralls speculated that smaller size gives the males of some species (whales, for instance) greater mobility in mating. So selection may not so much have made females large as males small—female choice again, but in these cases favoring agility, not bulk.

Perhaps also: compliance. In many of these species, females dominate males in disputes over food and other resources, despite males being more aggressive. In our close cousins the bonobos, females control male aggression through alliances, not larger size, although there are other matriarchal species in which females are smaller than males. Perhaps the best explanation is that larger females breed better. This is certainly true of many insects and fish, and there is evidence for it in mammals.

But Ralls also reached a surprising conclusion: "Once female mammals became committed to internal gestation and lactation, their parental investment was so great that the likelihood of evolving a social system in which the relative parental investment of males exceeded that of females and males became a limiting resource for females became exceedingly slim." In other words, once you go through pregnancy, you aren't going to act like a female jacana. Although females in these mammals loom over and dominate males, they aren't like jacanas, because even a doting mammal father is going to have a tough time catching up to the investment his mate has had to make before their young were born, as well as the ongoing investment she is committed to biologically in producing milk. When the platypus genome was sequenced, this primitive egg-laying mammal turned out to have genes for three milk proteins matching our own, despite the fact that the platypus mom doesn't even have nipples but secretes her milky fluid from modified sweat glands spread over her chest and belly. This means milk is as old as mammals—over 200 million years, to the common ancestor between us and the platypus—and later evolution did not cut back on mothering. Yet as we will see, some South American monkeys have fathers that invest so much they actually give mothers a run for their money.

Since Ralls's pioneering work, we have learned a lot more about species in which females dominate males. She knew of some evidence that smaller males are favored in birds of prey because of greater hunting agility, and this has now been proved. Oliver Krüger's sophisticated comparison of 237 species in the three main groups—falcons, hawks, and owls—failed to find a role for female competition or male agility in courtship, but it did strongly support the hunting hypothesis: in species that hunt more agile, rarer, and larger prey, males are smaller than females, because in these birds the little nimble males provide most of the females' food while they incubate eggs and brood chicks. As Ralls thought, reproductive demands can keep females large.

Her work on mammals has also been brought up to date. A comprehensive 2011 review by Patrik Lindenfors and Birgitta Tullberg concluded that almost half of all mammals have males that are larger than females, "a pattern that is clearly linked to sexual selection." Many fewer mammals have larger females, and there is little or no evidence that this reversal is due to sexual selection à la jacanas. So half of mammals have larger males, many have males and females who are the same size, and some have larger females, not because of competition for access to multiple fathers but because large size enables you to breed better in the ancient mammalian way. It also means you can defend resources, including your own body, against male demands. But that doesn't mean females don't compete for males.

Ralls looked briefly at hyenas, and new research has shown them to be among the most fascinating of mammals, completely belying their negative reputation. Zoologist Kay Holekamp and her colleagues, looking back on their twenty-three years of research on Kenya's spotted hyena in 2011, noted the great complexity and high level of cooperation in hyena society, including dominance hierarchies and coalitions resembling those of baboons. This favored unusual traits: "Adult females are larger and more aggressive than adult males, they are socially dominant to all adult males born elsewhere, and the female's genitalia are heavily 'masculinized.' These unusual traits not only give females top priority of access to food, but they also give females virtually complete control over mating." In the popular view, hyenas are nasty scavengers, while lions are glorious big-game hunters. In fact, hyenas, too, are top predators (killing 95 percent of their food, including the occasional human), and lions probably scavenge more than hyenas do. The two species often fight over a fresh carcass, whoever killed it. They have long shared habitats, as indicated by the thirty-thousand-year-old cave art of Lascaux and Chauvet.

However, females have very different roles in the two species.

Lionesses are smaller than lions and don't sport gorgeous manes, but they do most of the hunting, often in small groups with their sisters. Males lounge around and eat—they are happy to take charge of a carcass killed by females—until strange males show up from out of the wild, in which case they will have to defend not only their life of leisure but also the cubs they have sired with the pride's females. Foreign males will try to kill the cubs (over the usually vain protests of their mothers), and if the fathers can't protect them either, those fathers will be driven out, often doomed, like the infants, by deadly wounds. In due course, the lionesses will mate and breed with the new males. Manes, by the way, cause overheating when males hunt, but females nevertheless prefer the darkest and densest tresses. Males do feed their cubs on the meat they get, but it's probably stolen from females or hyenas.

Hyenas, in contrast, live in a matriarchal world. Their large groups are collections of maternal kin, and female hierarchies are critically important. There is up to a fivefold difference in reproductive success between high-ranking females (who live longer, reproduce earlier, and have more surviving cubs) and their low-ranking counterparts, yet the least dominant female trumps the most dominant male. Males stay or leave on the basis of female choice, and when a male approaches a female in estrus to court her, biologically "ready" as she is, she may not be ready for *him,* so he is risking life and limb on his fond hopes. Both sexes are promiscuous, but the male is the supplicant and the female gets what she wants. Micaela Szykman and her colleagues closely observed many matings. In one, they reported,

> Assumption of the receptive stance by the female appeared to signal the male that it was safe to mount her. After several pre-mounts, the male mounted the female, and repeatedly attempted to achieve intromission. This task was apparently made extremely difficult by the female's peculiar genital morphology. The male had to squat down and under the female, so low that his rump was sometimes on the ground, to maneuver his erect penis into the female's flaccid phallus.

Phallus? Or else call it a clitoris that prenatal hormones have hugely enlarged. The report explains further:

> There is no external vagina, as the labia are fused to form a pseudoscrotum, and the clitoris is elongated and fully erectile such that it strongly resembles the male's penis in size and structure. This pseudopenis is traversed to its tip by a central urogenital canal, through which the female urinates, copulates and gives birth. This unique

> female morphology makes intromission by the male considerably more difficult than it is in other mammals and also makes copulation by force physically impossible. Although [she] retracts her clitoris into the abdomen to permit penetration by the male's penis . . . the male hyena typically experiences unusual difficulty locating this opening and achieving intromission.

So the needy, fearful, disadvantaged males "must overcome unique motivational challenges associated with approaching and courting large, aggressive, well-armed females." Female hyenas are *more* "masculine" than males in size, dominance, and fighting ability, *as* masculine in their genitalia, and yet bring life into the world and nurse their young. Males not only have to be fully approved to offer their dollop of genes, they often have to slide under the female, and rape is anatomically impossible.

These voyages around the world to look at how creatures great and small arrange their love lives are meant to help us think about our own historically fraught ways of doing things and, especially, to make us realize that male domination, however common, is not the only way and is no more natural than female dominance or, for that matter, equality. But there are very few systems where sex does not involve female choice. And today we can delve much more deeply into that pivotal process than was ever before possible.

The two most exciting trends in biology today are the close study of functioning brains and the rise of gene science. When you put them together to find out just how genes make brains make behavior, the thrill is doubled. Add to that the possibility of focusing these tools on closely related species that act quite differently, and you open new worlds of discovery.

Consider the small tropical fish studied by Molly Cummings of the University of Texas at Austin, described in her 2012 paper "Looking for Sexual Selection in the Female Brain." She began with a pair of closely related fish species that differ greatly in female choice: one has a lot of it, the other very little. Both have internal fertilization, which means the male must achieve physical intimacy with the female, and both bear live young, called fry.

The female-choice species is the El Abra pygmy swordtail, found in Mexico's Río Pánuco. It's a two-inch-long, sleek fish with dull greenish-brown scales, except for a black streak flanked by two rows of shiny blue stripes like arrow feathers. The male adds a long, thin black swordlike strand streaming along the bottom edge of an otherwise translucent tail fin. The male-coercion species, called the mosquito fish because it feasts on those insects' larvae, lives in the Mississippi basin. It's an iridescent yellowish-gray and about the size of the swordtail, except that the females are quite a bit larger than the males.

Cummings separated the sexes with a glass partition and let females stay as near a male as they liked. Swordtail males have a range of sizes; the smaller are less often chosen and (oddly enough) more coercive. Perhaps coercion stems from their size disadvantage, as it sometimes does in insecure human males. Females were exposed for half an hour to one of four choices: simple (between a large and a small male); minimal (two small males); same-sex (two size-matched females); and two empty compartments. Looking at the gene activity in their brains after this, Cummings zeroed in on four genes that turn on in the brain only during female choice, including in the females who chose between two small males.

Repeating all this with the mosquito fish, she did find some weaker preference for larger males, but the genes that were more active in female choice in swordtails were *less* active in female mosquito fish, which in the wild actually have *less* choice and *more* male coercion. Cummings reasoned that mosquito fish females might actually have to suppress their choosy brain genes to allow the rough mating process in their species to unfold—regrettable, but in their case reproductive.

The brain genes in question are found in many animals and have to do with remodeling brain circuits, one basis for learning and memory. Female choice in swordtails may be partly learned through experience with males, while mosquito fish may have to suppress the same learning process. Cummings is now trying to test this further by knocking down and overexpressing brain genes and then watching the females make their choices. The possibilities for transforming sex and courtship are mind-boggling.

Cummings was partly inspired by another comparison between closely related species that mate differently: the voles. The vole studies have been going on for over a quarter century—they were begun by Sue Carter and carried forward by Thomas Insel, Larry Young, and many others. First, Carter and her colleagues showed the importance of the now-famous brain chemical oxytocin in mothering. This provoked countless studies of oxytocin in attachment, friendship, and trust; even humans, when exposed to an oxytocin nasal spray that quickly reached the brain, became more sociable and trusting. Oxytocin has also been tried on people affected by autism, which is in large part a problem with social bonds. This ever-growing body of research was reviewed by Carter in 2014.

Oxytocin has created an exciting frontier in autism research, but it has also unveiled the basic biology of bonding. After working with Carter, Insel began comparing prairie voles, which are pair-bonding and paternal, with montane voles, which are multiple mating and only maternal. Male fidelity and fatherhood turned out to depend on a related brain chemical, vasopressin, but it was not the *level* of it that mattered—you could inject montane males with it and even then they wouldn't commit. What counted was the location of receptors for it in the brain. That, in turn, depended on gene expression; the genes that made the receptors were switched on in key emotional circuits in the brains of prairie voles but not montane males. The switch is in the promoter for the gene, which can be long or short. In 2013 Zoe Donaldson and Young showed how the gene controls the expression of the receptor in the limbic system, also known as the emotional brain.

Now comes the astonishing part. Because we know a lot about how to manipulate mouse genes, and because normal mice are neither pair-bonding nor paternal, Young's lab put the prairie vole gene promoter (the long version) into male mice, and these remade males now wanted to hang out near females. Other vole species also fit the pattern: pine voles are pair-bonding and paternal, and the males match the prairie vole pattern of brain genes; meadow voles mimic the montane voles' brains and behaviors. Yet Miranda Lim and her colleagues, doing the same kind of prairie vole gene insertion that Young had done with mice, made wayward meadow vole males into new men, in the prairie vole vein of loyalty and fatherhood. Their new genes made them family guys.

And not much of a gene change is needed—which helps explain why such similar species have such different mating habits. If some change in the environment gives pairing off and fathering a new advantage, evolution can almost flip a switch and produce (over many generations) a new mating system. And all this doesn't just apply to voles. At present, Donaldson is exploring the same gene in primates to find out how all this works in some of our closest relatives. In chimpanzees, studied by William Hopkins, Donaldson, and Young, males with the short variant of the gene promoter were considered less dominant and more stable in their personalities than females. In 2014, Stephanie Anestis and her colleagues reported the first research on the gene in wild chimpanzees; they compared the wild population in the Kibale National Park, in Uganda, with those at the New Iberia Research Center, in Louisiana, and the chimps at the Yerkes primate center that Hopkins's group had studied.

The long promoter was much more common in the wild East African chimps than in either captive center, where the chimps descended from West African forebears. In the males in the Louisiana population, the long promoter was associated with "smart" social behavior, such as forming coalitions and initiating play, as well as with friendliness; this does not seem to fit with the Yerkes finding of more stable personalities and less dominance in males with the short version, but the behavioral measures were very different. The needed behavioral analyses of the wild chimps have yet to be done.

However, all the studies so far point to the importance of the vasopressin promoter as a basis for variation in social behavior. And most remarkably of all, research in Sweden by Hasse Walum and his colleagues has shown that natural variation in the promoter (although not the same as the variants in voles) is linked with partner loyalty in humans.

Sometimes after an expert delivers a lecture on this research, a young woman will come up and ask, half-jokingly, for something she can put in her boyfriend's beer to make him commit. There is no such thing now, because we can't insert genes through beer; if your boyfriend uses snuff or an inhaler, you could theoretically add oxytocin, which seems to make people of both sexes more trusting and affectionate. (Don't try this at home.) But the question of whether we as a species could evolve a more perfect marital union, with more loyal males who are better fathers, is something else again. As we will see, we may well be in the midst of such an evolutionary change right now. And who knows: perhaps the next generation of scientists will find a way to speed up the change by spiking beer.

## Chapter 4

✦

# Primate Possibilities

Now we come closer to our own corner of the biological world and look at the roles of males and females in our cousins, the primates. In the evolutionary time scale we began as primates called prosimians, and so they have much to teach us about where we have been and where we are going. And they have news for us about male supremacy.

Lemurs, the most common prosimians, dwell—or, rather, move restlessly—in the lush forests of the large island of Madagascar, off the east coast of southern Africa. They are less like us than monkeys and apes but more like us than any other mammals. Once found throughout the world, they were beaten back by the monkeys their ancestors gave rise to—except in their island world, which monkeys never inhabited. They represent roughly the kinds of animals we were before we became monkeys or apes, and that's a big part of their interest.

But there is also this: in all lemur species, females dominate males. This happens in the wild and in captivity, in one-on-one fights over food, turf, and elbow room, in subtle and crude ways.

As Alison Jolly—the great pioneer of lemur field studies who died in 2014—put it, lemur life is one long soap opera. The males have nasty fights among themselves—over females, for instance—and females compete with one another as well. In some but not all species, females are larger. But the key fact is, females dominate, while cowed, submissive males know and keep their place. Females can dominate one-on-one, but they usually do it through female coalitions as well. This is called the *lemur syndrome,* and it may be a key to the evolution of primate social intelligence.

Take the blue-eyed black lemurs, stunningly beautiful animals with striking eyes, from pale azure to a shocking electric blue, set in plush black fur. As they meander through the trees, balancing on the branches using their long, lithe, bushy tails, they might almost seem like large, elegant cats, but their eyes are deeply intelligent and their hands look and work like ours. In a study by primatologists Leslie Digby and Alexandra McLean Stevens, blue-eyed black lemur females won 99 percent of contests with males, combining social coercion—threats and subtler messages—with more frequent outright aggression. Growing up, the boys did win a few contests, but girls still ruled and got better at it as they became unbeatable adults. Despite their submissiveness, some males manage to mate with multiple females, which means that there is serious female choice.

Yet varied lemur mating patterns all seem to work in the female's favor. Brigitte Marolf and her colleagues compared the red-bellied lemur, which is monogamous, to the crowned lemur, where both sexes mate multiply. Crowned males were more aggressive with each other, but females were also tougher on them, and they dutifully groomed females more than did their faithful red-bellied counterparts. Nevertheless, despite these differences, males in both species bowed to female power.

This is also true in ring-tailed lemurs—gray creatures with striking black bands circling their long, thick tails—who live a highly social life. They like to huddle and sunbathe, bellies angled skyward, which makes them seem mellow, but every female dominates every male, despite the fact that the males are just as big. They also have similar androgen levels, although females double their levels during breeding season—along with their belligerence. Psychologist Christine Drea and anatomist Anne Weil showed in 2008 that (like hyenas) ring-tailed females have naturally male-like private parts; from a distance these organs could be a penis and testicles, and when measured up close they have the anatomy of female primates given male hormones. Their ultimate ornament is an enlarged, dangling clitoris—as in humans, the only organ that has no known function except pleasure, but in the lemurs it is much larger. So ring-tailed females are not just lording it over males but also probably thoroughly enjoying it.

Yet in the end, the cat's meow of female satisfaction may occur in the minute gray mouse lemur. Both sexes are wide-eyed cuties, Yoda dolls straight out of a *Star Wars* collection, but like teacup Chihuahua puppies, they will fit in the palm of your hand. Yet these females too have no trouble lording—ladying?—it over the pliable males in their nocturnal world. One research paper, titled "Sex in the Dark," showed that males must travel widely to reach their hearts' desire, but with females calling the shots, most litters have multiple fathers. In another study, published in 2012, ecologist Doris Gomez and her colleagues found that females prefer males who defeat other males, despite the fact that those same winners can't beat *them.* Picture perhaps an athletic princess who watches her suitors wrestle it out (knowing she could pin the best of them) and then deigns to let the winner do her bidding—on her terms.

In a 2012 laboratory study, behavioral ecologist Elise Huchard and her colleagues controlled the size of the sexes. She predicted that smaller females would mate with more males—"convenience polyandry," it's called—since they'd be less physically capable of resisting the boys' advances. Wrong. The larger the female, the more mates; the researchers called this "adaptive polyandry," because those females give the nod to one after another—the wrestling matches just decide who's first in line. The paper drily noted, "Mouse lemur females exert tight control over mating and actively seek multiple mates, suggesting that polyandry might constitute a more rewarding strategy" than the usual pattern of female choice *before* mating. Rewarding, perhaps, in more ways than one.

The great primatologist and evolutionary anthropologist Sarah Blaffer Hrdy gave us glimpses of all this decades ago, in a book intriguingly called *The Woman That Never Evolved.* That nonexistent female was the passive one of male myth, a sex object mysteriously both cool and hot, out of reach and responsive, virgin and courtesan, with nary a tough bone in her body or an aggressive thought in her head, who magically morphed into an all-giving earth mother as soon as you—you clever boy—figured out how to bring her down off her pedestal. This myth had an evolutionary version, in which female primates stood by while males fought over them, winner take all. What could the poor delicate things do, anyway? They made the best of it. After all, didn't they just want the best fighters to protect them, the best hunters to feed them and their babies, the boldest, toughest sires for their sons?

The myth ran deep in our culture. It was Achilles and Agamemnon fighting over the beautiful slave Briseis; Greece and Troy spilling rivers of blood over lovely Helen; Odysseus, after countless exploits and no few dalliances, come home to find Penelope, after twenty years, loyal as ever, warm, waiting, having rebuffed his every rival. And if you think the Judeo-Christian tradition is different, read Genesis or Judges. In these sacred sagas, men fight, while with few exceptions women wait and take what they can get. At best, women have wiles, only rarely pluck or strength. But these, we will see, were accounts of a very different type of culture than the one we evolved in. And women, in all likelihood, did not write those books. The woman that *did* evolve was a different one.

Hrdy knew this because she was a courageous field primatologist who braved many difficulties to do a landmark study of langurs—large, sleek, grayish-tan monkeys of central India. She made famous a phenomenon that intellectually tacked down one end of a continuum of extreme male violence: competitive infanticide. Others would find it in many other species—over fifty at last count among primates alone. But her langurs were a stunning first revelation. In a nutshell: Their troops had unstable male hierarchies. Females, their relatives, and their young formed the core, but the males would, after a couple of years in power, be driven out by other males. When this happened, the new males killed all infants under six months of age, which made their victims' mothers sexually acquiescent again—this time, to them.

At first glance, it seemed the old story of pure and rather vicious male privilege. Females had to watch their infants bitten and torn and, in due course, had to mate with their tormentors, much as the women taken in slavery in countless wars became wives and concubines of their enemies. But, in fact, it was not so simple. Mothers fought back. Langur females valiantly tried to prevent the new, empowered males from getting to their babies and risked their own lives trying to protect them, holding on or grabbing them back even after the infants were gravely wounded. And they helped each other, deploying alliances of sisters, mothers, and female cousins, sharing the risk and cost of joint resistance. If they succumbed in the end to overwhelming force, it was not without a fight.

The old Western myth was wrong in many other ways, too. Female primates were never passive in any situation. Even in ordinary times, when they weren't facing anything as intense as infanticide, they were resisting male abuses. They were constantly active on their own and their kids' behalf, jockeying for position in a female dominance order, competing over males, food, and territory, shielding their young from harm and trying to ensure that they would grow up into roles where (male or female) they would be dominant, too. If some primate females had to live in a male world—and not all did—they would not just stand by; they would make the most of it.

In time, Hrdy realized that making the most of it in sex meant trying to exercise choice not just before mating, which was indeed often possible, but after mating as well. Like the diminutive gray mouse lemur, all primate females had options: mating opportunities beyond the first; "extra-pair" copulations (where there was a pair); adaptive or convenience polyandry; sex with other females; and, in the right circumstances, selective promiscuity. Even in pair-bonding species, where both parents contribute to offspring care, your chosen boy can get gobbled by a predator, fall out of a tree, be chased off by other males, or just move on to greener treetops. A girl can't be too careful—she needs sexual insurance. And she also needs the best male genes her own sex appeal can buy. Growing evidence on all these points, throughout the animal world, led Hrdy to write a famous paper called "The Optimal Number of Fathers." At this point you won't be shocked to learn that the optimal number often isn't one.

Not just in primates but also across mammals, the most impressive fathers are our not-too-distant relatives the small South American monkeys called marmosets and tamarinds. They live in the forest in small (although not always conventional) families in which fathers are permanent and do a wealth of parenting. Except for occasional twins, having one tot at a time is a primate hallmark, enabling a long, slow course of parenting and learning. But marmosets give birth to twins routinely, a feat possible mainly because of fathers.

It's no easy task to keep twins alive in nature. Even in human populations, before modern times, fewer children survived from one hundred pairs of twins—that's two hundred babies—than from one hundred single births. That is, twins died more than twice as often as singletons, which explains why human twinning was historically rare. Selection worked against it, despite the apparent advantage of doubling your progeny.

But along the evolutionary way, marmosets stumbled on a trick to get around this: be the most involved fathers in the whole mammal world. They can't get pregnant or breast-feed—even *they* couldn't evolve a way out of 200 million years of biology—but they do everything else. In their thick, high forest realm, where infants would fall to their deaths or be picked off by a hawk or snake if left alone for a few minutes, they aren't—not even for seconds—because anytime the twins are not suckling on their mother, they are riding dad. That's twenty-four/seven, except for short milky interludes, carrying twins whose combined weight is 20 percent of his and counting, while in the daylight hours constantly moving through the trees and trying to find enough food to stay alive and pay the huge caloric bill.

Now, the marmoset mom's life is no picnic, either. She has to find food for three, after carrying the twins inside her for months and enduring the risk of delivery. They're 20 percent of her size, too; it's as if *we* routinely gave birth to two fifteen-pounders, who proceeded to grow apace and breast-feed to their hearts' content. But of the dozens of monkey species around the world, only these few can afford twins consistently, because the burden is as shared as it is in any pair-bonding bird. And although Mom and Dad are not winging it to and fro between food sites and a nest, they are bearing a big load. In fact, in some species of marmosets, families are made up not just of two parents but helpers as well—semi- or even fully adult males or females who refrain from reproduction while caring for the main pair's young.

Consider a pair of golden-white tassel-ear marmosets just before the female gives birth, a few steps ahead of a stalking jaguar that smells pregnancy in the night. After some failed attempts, she finds a tree with a large hollow not engaged by a raccoon or other forest denizen also in need of shelter from jaguars. At sunset she goes into labor, with contractions every five minutes. She squats and stretches her body upward, convulsing a little and squeezing her eyes shut with each contraction. Her mate is nearby, waiting to assist as only marmoset males do. In less than an hour, the contractions begin to pulse quickly, and the first twin appears. She licks it as it emerges, and it soon clings to her chest. The male leans out of the hollow's portal into the night, checking for predators. In a few minutes, the second baby is born, the male attentively sniffing it as it comes out. Soon he is chewing the two umbilical cords, eating them like long strands of spaghetti. Then he gorges on the placenta as the female licks the twins.

The next day, the four emerge and resume the forest trek. The twins cling to their mom while her mate grooms her, but for now she rebuffs his bids to hold his young. Soon they are joined by others in their extended family who have spent the night elsewhere. All show intense interest, sniffing and nuzzling the young, until finally the father's prerogative is recognized, and the mom allows him to hold one. He cradles it tenderly. One young helper pulls the second twin off the mom but is soon in distress as she realizes it is clinging for dear life; her first babysitting gig ends as Mom relieves her. Within a few days, the father is carrying the twins almost all the time, giving them to the mother to nurse and to occasional helpers, who do help.

But this is not the case in most primates outside the marmoset family. Helpers in many species do little, and sometimes they do more harm than good. As for males, fathers or not (they often don't know), their contribution is usually minimal.

Barbara Smuts, another brilliant primatologist of Hrdy's generation, studied baboons on the African savanna. Not only are these ground-living monkeys much closer kin to us than lemurs or marmosets, they are highly successful—they're considered vermin in some countries, a backhanded tribute to their adaptations—and they are large, dangerous, brainy primates with intricate social lives.

As such they're a glimpse into how our remote ancestors may have adapted to life on the ground. Irven DeVore, a mentor of Smuts and Hrdy, had shown that in the wild, baboon males ruled in their very large troops—up to hundreds together—not by force alone but through alliances. (We did not know it then, but alliances were also common among females of many species.) You could figure out a dominance ladder by tossing a chunk of food between any two males. The results of all possible pair contests would then tell you which male could beat every other one-on-one, which could beat every other except the first, and so on down the pecking order—a term derived from birds, where it was all about who pecked and who *got* pecked.

But in these smart monkeys, that ladder from the experiment didn't tell you who was in charge in real life, because baboon dominance demanded coalitions. It wasn't whether Chuck could beat Sam in a one-on-one bout; it was whether Sam could rely on Joe and maybe Fred to join in. Chuck might be the strongest and most vicious, but if he couldn't assemble an alliance, each of them would trump him as long as they didn't get caught with him alone. Coalitions also helped protect the group from big cats and other African ground predators. Mothers with infants would stay in the center, juveniles would be nearby, other, weaker adults would gravitate toward them, and at the periphery teams of the biggest, strongest males would face down the threat.

In a species that grew up slowly, with much to learn about social life—friends, foes, faces, threats, coalitions, hierarchies—the mother-infant relationship gradually ushered the youngster into a fairly safe place in a complex social world. Play, too, had a major socializing role, as infants and juveniles tumbled energetically over one another. But baboons had to grow up male and female, did so at different rates, and ended up different sizes. Males were not only bigger and bulkier than females, they had much larger canine teeth, to scare and, at times, to tear.

They could threaten females, but they were especially hazardous to other males. Recall that in systems in which males can mate with many females—gorillas, for instance, where top males keep harems, or, in the extreme, elephant seals, where 4 percent of the males get 85 percent of the couplings—the biggest losers in evolution are not the put-upon females but the totally shut-out males. This is true to some degree in baboons, although as in many primates, there is a time dimension; males age out of power and breeding. While this might seem a rotation that in time gives access to all, males have to age *into* mate-worthy positions, and in many species this means male transfer at adolescence. When female kin are the core of the group, young males are pushed out as they near adulthood, and this is a dangerous time for them. Mortality is high. To survive well, and certainly to mate, they must be admitted to other troops, where the stronger, older, well-connected, strange males don't want them, any more than their dads and uncles did back in their home troops.

Smuts discovered a marvelous workaround used by young male baboons trying to break into such a brave new world. They couldn't get close to sexually receptive females, who were monopolized by dangerous resident males, but they were able slowly to get to know females with infants. In her remarkable book *Sex and Friendship in Baboons,* Smuts described the pattern in quantitative detail. It's the opposite of the langur case. There, the new males are young but strong adults, and they confront the older local males first. If they win, they go after infants with a vengeance and, in due course, mate with their mothers. Baboon male strangers, in contrast, sidle up to moms with infants, hanging out at a safe distance with postures and facial expressions that say, "No threat." Over time they move closer, begin to groom the mother, and finally touch and hold the infant.

All the while, any aggression from the established males is dampened by the nearness of the infant. This "agonistic buffering" occurs in many primates, where weaker males or females protect themselves by borrowing infants. This is often dependable insurance against attack. But in many months of study, Smuts got to know her subjects well; she found that the females who befriended the young males were hedging their bets. Ultimately these males could become not just pals but protectors, and maybe even lovers. In evolutionary terms it was a win-win; males were gaining tolerance and gradual acceptance in what could have been a deadly new troop, while the females were optimizing the number of fathers. If these were friends with benefits, the benefits were mutual and might last a lifetime.

So even when dominated by larger males, females can gain advantages in survival, offspring protection, and mate choice—including multiple mates—by subtly deploying power over time. Many *female* alliances achieve similar ends in the face of male threats; but here was a female-male alliance based on a young mother's tolerance and indulgence of a new, vulnerable, supplicant male.

Smuts knew, though, that this rather sweet and, for females, empowering process was not the only way baboon males approached females; scattered throughout the primate world were much nastier patterns. In 1993, she and her father, Robert Smuts, wrote a paper on male sexual force in primates that was as much of a turning point as Hrdy's work on infanticide. Such coercion, the father and daughter found, was widespread in primates and mammals generally. This was not unknown, but their systematic overview stimulated a great deal of further research, much of it gathered in a 2009 book edited by primatologists Martin Muller and Richard Wrangham. Some male coercion has a short-term goal, but some results in long-term relationships; some seems designed only to persuade, while some results in very serious injury.

Reports by other primatologists are ongoing. Nicole Gibson and her colleagues recently published a case of rape among Peruvian spider monkeys. We know them as graceful acrobats that seem to fly through the forest using their long tails as a fifth grasping limb. But in one less-than-graceful case a female was set upon by a male who "approached fast, crashing through the canopy." She escaped, but he chased and attacked her stubbornly for ten minutes. The fight involved "cuffing"—hitting her on the head repeatedly with his arm—"and fierce wrestling." She screamed and whined; he growled. She "dropped down to a low vine next to the tree trunk, but [he] pinned her to the vine and started to copulate." She defecated twice, something females never do during normal sex but often do when attacked. She scrambled up the tree, but he quickly caught and entered her again. He ejaculated and stayed in her for two minutes, until she finally squealed and jumped away from him.

Such forcing is unusual for spider monkeys, and the perpetrators are likely to be less successful males. Importantly—and in all species with male sexual coercion—females have ways of avoiding it. One is to form consortships with males they like, perhaps going off into the forest away from others for a period of days to mate privately. Or, when they are sexually receptive and mate in a group setting, they may immediately mate with other males, confusing them all about paternity and thus protecting their infants from infanticide. Here as throughout primate evolution (our own included) females have had options, but through no fault of their own, the options didn't always work.

In orangutans—the great apes of Sumatra and Borneo—forced copulation appears to be more like the rule than the exception. A female and her young live mainly as a single-parent family until a male shows up. She prefers older, larger, more experienced males, but as she ranges through the forest she may encounter younger, more aggressive ones. These oversexed, frustrated, bullying subadults may not take no for an answer, and although they may be only half the size of mature males, they can gang up on a victim and overpower her. She's most at risk when, after years of nursing, she begins to be fertile again.

ElizaBeth Fox wanted to know what females do about it. She spent nine thousand hours over two years watching wild Sumatran orangs and saw more than 200 copulations. Of the 141 that involved subadult males, 99 percent were male-initiated, and females resisted at least a third of them. Orangutan females try to keep harassment to a minimum by hanging out with mature males—the only ones *they* initiate sex with. Yet these associations are not always sexual, at least not in the short term. Females hang with older males for protection, and this keeps thuggish teens at bay. In 2010, Cheryl Knott and her colleagues studied Borneo orangs, refining Fox's work. By analyzing the hormones in the orangs' urine, they knew when the females were fertile, and this was when females actively sought mature males. The females couldn't avoid unwelcome advances at other times, but resistance was usually futile anyway. It was a good way for females to confuse young males about who the father was and reduce the chance of infanticide by these same bullies. Meanwhile, they got a fair chance of choosing the father they wanted.

This also turns out to explain female toleration of male bullying in other primates with larger and stronger males. Interestingly, Joyce Parga and Amy Henry reported on an event among our old friends the female-foremost ring-tailed lemurs—a rare case of sexual bullying that, paradoxically, proved the rule. A male tried to force himself on a pubertal female having her very first estrus, achieving penetration despite her struggles against him. She showed her preference for other males by presenting to them, but she rejected this one—a known loser with very low mating success. Parga and Henry think his behavior, rare in ringtails, was due to human food provisioning (the group was not completely wild), which made this young female come into estrus early; she had not yet grown into full command over males. So the ringtails keep their reputation of female rule, despite a dud male forcing the equivalent of a young teen. She no doubt went on to learn how to keep such fellows in line.

If you're getting the idea that females dominate in primate studies almost as much as in lemurs, you are not far off, at least for Western primatology. Jane Goodall, Dian Fossey, and Biruté Galdikas largely put the field on the map in the 1960s with classic studies of chimps, gorillas, and orangutans. Among other outstanding fieldworkers who followed, aside from Hrdy and Smuts, were Jeanne Altmann, Shirley Strum, and Joan Silk on baboons, Anne Pusey on chimps, Dorothy Cheney on vervet monkeys, Karen Strier on muriquis, Linda Marie Fedigan and Katherine MacKinnon on capuchin monkeys, Meredith Small on macaques, Alison Jolly, Alison Richard, Patricia Wright, and Patricia Whitten on lemurs, and many others—not to mention Sue Savage-Rumbaugh, whose work on the mental capabilities of captive chimps and bonobos transformed our ideas about language and the mind. As for the coming generation, it is estimated that four out of five Ph.D.s under way in primatology will go to women. It's been said that primatology done by women is inherently feminist, but these women's goal was to advance science—although they may well have noticed things about sex roles that men missed. It's not that men haven't mattered to the field; they have—Japanese primatology, in particular, has remained a mostly male domain, and other men have done important primate fieldwork. But women have made an exceptional impact from the outset.

None more than Jane Goodall—now a conservationist and cultural icon—whose bold and brilliant research on the chimpanzees of Tanzania's Gombe Stream Reserve during half a century made everyone think anew about the boundary between us and animals. She described their individuality, sensitivity, depth of feeling, relationships and communities, devoted mothering, lifelong family ties, respect for elders, ability to make and use tools and pass on traditions, cooperation during hunts and meat-sharing afterward, and many other aspects of their lives that seemed, well, human. But as she later admitted, her initial picture of chimps made them seem a pristine and nearly idyllic version of almost-human life. *In the Shadow of Man* was the title of her first book about them, and it emphasized the vulnerability of these apparently gentle creatures in the face of human violence and greed.

Years later, however, writing her scientific masterwork, *The Chimpanzees of Gombe,* she reported unpleasant findings that made them seem anything but gentle. Two in particular stood out. First, there were two females, a mother and her grown daughter, who took to killing other females' infants. They did this not once or twice but many times, as if an obsessive and grim pathology had taken hold of them. This was not like the langurs, in which foreign males enhanced their reproductive success by killing the infants of other males and later impregnating the females. But it did resemble the patterns of infanticide in some species—wild dogs, for instance—in which females kill others' young, with the effect of making room and resources available for their own.

Equally disturbing and in the end more common was a pattern now widely seen and studied in wild chimpanzees across Africa. A number of chimp males—a gang, really—patrols the edge of the group's territory in the forest. If in their travels they happen upon a member of a neighboring group, they promptly set upon that chimp en masse and beat, kick, bite, and stomp him to death. Over time, one group may eliminate another and take over the victims' territory. This habit was almost human in a very different sense, and it was not welcome news, coming from a species that is literally our next of kin; in DNA sequence, we are about 98 percent chimpanzee.

The pattern has been confirmed many times in different chimp populations. John Mitani, one of the leading field primatologists in the generation following Goodall, reviewed the findings in 2009. By then long-term studies had been in place not only in Gombe but in the Mahale Mountains (also in Tanzania), in the Budongo Forest and the Kibale National Park (both in Uganda), in Boussou in Guinea, and in the Taï Forest of Côte d'Ivoire. Kibale has long-term studies of two separate chimp communities, and more recent work has begun at Mount Assirik, in Senegal, and in the Republic of the Congo. So now we have decades of data on chimp groups in forests, mountains, dry savannas, and blends of these, from East to West Africa across thousands of miles. We have history. All of this information has revealed specific adaptations, variation that is both genetic and cultural, and yet also some persistent findings.

Competition between males, both within and between communities, is routine, partly because chimp females have obvious estrus, which makes what is called the operational sex ratio—the number of males who might want to mate with a *receptive* female at a given time—very high. Dominance is critical; by genetic evidence, a third to half of all infants are sired by the single male at the top during his tenure. Male coalitions make all the difference, and these alliances trump pure physical strength. The half or more of the mating opportunities that the alpha male doesn't get, he cedes to his coalition partners. Dominant coalitions have lasted up to seven years, punishing and pushing back challengers in the group. Eventually, as dominants age, coalitions can't keep command, and a revolutionary alliance may arise.

A second function of male coalitions is hunting, seen in several different forest chimp communities. Red colobus monkeys are typical prey, and chimp males do best when they strategically deploy to ambush a stray colobus in the trees. These hunts can seem deliberate and coordinated and can last five or six hours. Brave little colobus monkeys may mob the chimp males and even wound them, but hunts often succeed, and the meat is shared. This is a clue to the evolution of human sharing, since meat is the food most often shared among hunter-gatherers. The chimp version looks more like tolerated scrounging, and that may be where human sharing came from. Geza Teleki, Craig Stanford, and others proposed that male chimps would trade meat for sex, but in some environments they don't especially share with estrus females.

However, this may depend on local ecology or tradition. Cristina Gomes and Christophe Boesch reported in 2011 that in Côte d'Ivoire's Taï Forest, chimp males *do* trade meat for sex. In fact, meat for sex was one deal among several in a complex ape economy. In their sophisticated quantitative analysis, grooming and support in dominance challenges mattered, too. Grooming removes nits, which benefits the groomee and probably also releases endorphins, as a result of fur stroking. The groom*er* gets the tasty tidbit, but the benefits don't stop there, because grooming tends to be reciprocal: you stroke me and I'll stroke you soon enough. Also, chimps trade social support in frequent disputes. But they also trade *across* commodities. In particular, males exchange meat for support from other males, and females in estrus exchange sex for meat—they mate more with males that share it. If they didn't, females would almost never eat meat, and low-ranking males would have little chance of ever having sex.

So even in the man's world of chimp society, females influence which male gets to them. Rebecca Stumpf and Christophe Boesch showed that Taï Forest females exercise choice—meat sharing being one standard males have to meet. Focusing on the most fertile part of estrus, they found that active seduction—she may repeatedly offer her prized swollen, pink rump to her chosen beau-for-a-day—does affect who mates when. In almost a thousand sexual encounters, males were the initiators three out of four times. But of these male attempts, females resisted nearly three in ten and mostly succeeded in putting off the male's advances. In the quarter of encounters when *females* did the seducing, males gave in to temptation eight out of ten times. Females also get to say when sex ends, allowing longer copulations with males they like; in a promiscuous mating system, where males have evolved huge testicles to trump the sperm of rival males, this female stopwatch can make all the difference.

But if chimps are our close relatives, what does all this bad male behavior say about us? Before we answer that, we have another, very different cousin equally close by. That's the bonobo, and if we are roughly 98 percent chimpanzee, we are also 98 percent bonobo.

How did we end up with two different species that are both our next of kin? Well, *their* ancestor split from *us* around seven million years ago, but they've been separate from each other for only a million or two—since the Congo River formed and separated what we might call the chimpobos from the bonozees. Being poor swimmers, the two populations said their good-byes and in due course became chimps and bonobos. But after that short one or two million years, they were dramatically different in their behavior. So how do we sort out which is the right mirror for us?

It helps greatly that in 2012 Kay Prüfer and his colleagues sequenced the bonobo genome. We had the human genome in 2000, the chimp in 2005. Now that we have all three, geneticists can triangulate the comparisons and find out just which genes chimps and bonobos don't share with each other and which we humans do or don't share with either or both. But this will take time. In the meanwhile, ongoing research in African habitats is telling us a huge amount about behavior.

There are many similarities. Chimpanzees and bonobos look like relatives and overlap greatly in body size, although bonobos don't get as large as the largest male chimps. Bonobos have darker fur, puffy around the head, and the head itself is smaller. As Adrienne Zihlman showed decades ago, the relative length of the limbs—shorter arms, longer legs—is more humanlike in bonobos. Anatomically those are the main differences. But in behavior and reproduction, it's a whole other story.

Bonobos remain confined below the Congo River's big bend, where (because the most dangerous species is us, with our endless wars) they have not been easy to study in the wild. Yet this research has been done for many years, and it has been an ongoing revelation. In a sense, you could take chimp sex and violence and flip them. "Make love, not war" became a cliché about bonobos. It's a stretch to say "love" about an ape's subjective experience, but whether it's love, lust, or a mellow blend, it's clearly female-forward. It's almost as if the lemur syndrome somehow got from Madagascar to the Congo, leaping over the male chauvinist societies of baboons and chimpanzees to inspire this isolated population south of the Congo River, whose members are living lives we humans might aspire to.

Consider: bonobo females are not exactly dominant over males, but they make alliances themselves that give any would-be uppity male long pause before trying to dominate *them*. These coalitions may once have functioned the way the cabals of male baboons do, keeping the most aggressive, especially younger boys, in line. In any case, what we now see in bonobos is a much lower level of male aggression against females *and* fewer violent conflicts with other males. Not for them a periodic, gang-style, testosterone-charged patrol into the forest looking for hapless victims from a neighboring group. Not for them, either, mortal fights among themselves over access to females in estrus. And certainly not for them: coercing females into sex.

We didn't need the genome to understand that different physiology helped make all of this possible. Female bonobos do have estrus, but compared to chimps' it is anatomically less dramatic and behaviorally more spread out. They have sexual swelling, when they are more likely fertile, but it lasts longer and is mimicked by pseudoestrus—swelling without fertility—during pregnancy and the postpartum period. Even in real estrus, maximum swelling does not always cue ovulation. On a graph, female sexual behavior (such as soliciting males) seems smeared across many more days of the month in bonobos than in chimps. So males have much less of a signal as to when ovulation really occurs and much less incentive to provoke fights. They also get more sex, as several females are likely receptive on any given day.

But here is where bonobo females get the widest choice of all: when they incline toward sex, they don't have to choose males. They can and often do choose each other. Vulva rubbing is a common and favored pastime among females; sights and sounds go along with it that seem to denote pleasure and may reflect a kind of orgasm. Female bonds are strengthened by this intimacy—"friends with benefits" seems apt—and wider alliances are enhanced. Since (as in chimps) it's the females that change groups after puberty, these trysts help build strong relationships among females who may initially be strangers.

But males don't suffer as a result. Apparently there is enough desire to go around, and males can have plenty of leisurely sex with the same females. They occasionally have to fight or be on the alert, but they don't have to get seriously hurt or killed, and they don't have to languish in frustration while bigger, tougher males shove them aside. They just have to eat, drink, and be sexual. In fact, unlike other apes, bonobos combine the pleasures of food and sex, much as humans do. And they are the only apes that routinely have face-to-face intercourse, sometimes complete with broad, gratified-seeming smiles. For bonobos, the battle of the sexes is win-win. In addition to more sex and less violence, bonobos play in adulthood, something rare in chimpanzees. That makes them appear more human, since adult play has always been viewed as one of our best features. Finally, bonobos are altruistic in ways that chimps are not—ways that have also in the past seemed uniquely human; they share for the sake of being sociable, even with strangers.

If this portrait of bonobo society sounds too good to be true, perhaps it is—but only to this extent: all animal species have some conflict, and in every species sex is to some extent a cause. Compared to chimps, bonobos really are the "love, not war" species, and female coalitions get most of the credit. Nevertheless, they do have conflicts.

Gottfried Hohmann and Barbara Fruth studied wild bonobos in the Democratic Republic of the Congo (DRC), with more than fifteen hundred hours of observation from 1993 to 1998. They were interested in conflicts related to mating, so they analyzed observations of mixed-sex groups, the most common kind. Defining aggression as pulling, slapping, hitting, and biting, they found 391 instances of conflict where the sex of both parties was known. Of these, 38 percent were between males, 26 percent were female-on-male, 23 percent were between females, and only 11 percent were males attacking females.

However, aggression was more likely on mating days, especially just before, during, and after mating. Both males and females who *started* conflicts had greater overall mating success than those who did not. Females often harassed other females' mating attempts, and the harassment revolved around certain preferred males. Male-on-female aggression was unusual and confined to random interactions, not friendships, even though mating was more likely in the latter. Finally, male aggression against females did indeed produce retaliation by female groups, often supported by other males.

Compared to bonobos, chimps have much more male-on-female aggression, much less female-on-male, and variable amounts of same-sex fighting, but chimp male-male fighting is far more likely to be severe. As for bonobo mating success, rank does matter for both sexes, but since there is little male coercion, female choice is effective, perhaps paramount. Male violence is mitigated by male-female relationships. Martin Surbeck and his colleagues, in a 2012 hormonal study of wild bonobos, found that male aggression increased around fertile females, but only low-ranking males had increased testosterone. High-ranking males were often in ongoing relationships with females, had more sex, and didn't need a testosterone boost to get it. The other times female bonobos tend to fight each other are, remarkably, contests over the dominance rank of their adult sons. So important is their influence in the bonobo social world that the next generation's allotment of male power depends on it.

But what about sex between females? Zanna Clay and her colleagues published studies in 2011 and 2012, based on more than one thousand hours of observations of semiwild bonobos in a DRC sanctuary. Clay studied all aspects of these erotic occasions but focused on copulation calls—high-frequency squeaks and screams. These were more likely with males but occurred during female sex, too. The 674 female genital contacts were not random: most were in pairs of low-ranking females, next between low- and high-ranking, and last (rarely) between high-ranking females. The squeaks and screams rang out in about one in five contacts, mostly given by low-ranking females, especially with high-ranking partners. Low-ranking females were much more active in same-sex trysts, but high-ranking ones were much more likely to initiate. And if an alpha female was watching (ape sex is usually public), this "audience effect" made the calls even more likely. So the call is not just something physical but reveals the excitement of certain social contexts—and it helps build key relationships. Newly arriving immigrants are sure to be low in rank, and these public intimacies are their entry point.

Bonobos are not only less violent than chimps, more female-forward, and much more varied and affectionate in their sexual habits; they are also more socially positive and altruistic, even toward strangers. But how do these differences come about? Victoria Wobber, Vanessa Woods, Elisabetta Palagi, and others have in the last few years begun to study their development and have found the patterns in infancy; the two species differ very early on, which means the divergences could be hardwired.

New studies have opened the era of neurobiology and genomics of these striking differences between species. James Rilling and his colleagues, beginning in 2012, compared chimpanzee and bonobo brains using diffusion tensor imaging (DTI). In this technique, vivid colorized images show the orientation and size of the brain's connections in three-dimensional space. Since the connections are white matter, DTI also indirectly reveals gray matter volume. Bonobos had more gray matter in limbic system areas related to emotion, in particular certain regions in the right brain known to be involved in social life—including sexual behavior, social play, tolerance, and empathy, all of which are more common in bonobos. Strong connections linking the amygdala, which is involved in aggression, and part of the frontal lobe, which controls impulses, as well as part of the cingulate cortex, could easily enable bonobos to suppress aggression and feel empathy. The researchers go so far as to suggest an empathy "deficit" in the chimp brain. Put more positively, bonobos have brains that are anatomically more adapted for social cognition, empathy, and impulse control.

Ideally, we will soon progress to studies of gene expression—like those we saw in voles and mosquito fish, but noninvasive—that will explain these anatomical differences while linking the circuits to behavior. We are not there yet, although a great stride forward was made in 2012, when the first bonobo genome was published. This was long awaited, not because of an exotic curiosity about an endangered ape but because the triangulation of the human genome with those of our two cousins will yield insights into human traits and their basis in genes and the brain.

Even the early analyses, however, have found interesting things. First, bonobos and chimps have separated only one or two million years ago, less time than we thought for their big differences to evolve. Second, at least 97 percent of all genes are shared by all three species. Third, 1.6 percent of the human genome is more closely related to the bonobo genome than to the chimpanzee genome, while a different 1.7 percent is more like that of the chimp. Finally, for more than 3 percent of our genes, we are more closely related to one or the other than the two of them are to each other. These discoveries are a first step toward answering the questions that have most intrigued us about them: Are we more like bonobos or like chimps? How much did the last common ancestor of the two of them resemble either as they are today? And what was the common ancestor of all three of us like? Along the way, the answers to these questions will help reveal the brain genes that have made us, happily, more like the bonobos in sex, play, and empathy and, unhappily, more like the chimps in individual violence and territorial aggression.

But what about male dominance and abuse of females? Which species are we more like in that? The answer turns out to hinge on where in human history you try to find it, and we will now explore that history.

## Chapter 5

✦

# Equal Origins?

Camille Paglia began her masterful book *Sexual Personae* with "In the beginning there was nature." She continued: "Sex is a subset to nature. Sex is the natural in man. Society is an artificial construction, a defense against nature's power. . . . Human beings are not nature's favorites. We are merely one of a multitude of species. . . . Nature has a master agenda we can only dimly know." We are trying here to know it a little. Having followed evolution up to our closest relatives, we next come to ourselves.

The first phase of our evolution, hunting and gathering, lasted until about 10,000 years ago. Since fully modern humans, in terms of body and brain, arose between 200,000 and 100,000 years ago, we have lived the vast majority of our time on this planet doing what the few remaining hunter-gatherers do. In fact, we were doing it before we became fully human—*Homo sapiens sapiens.* This pointed repetition of the Latin word for "wise" may seem like special pleading, but this is what we call ourselves. This means you and me, but it also means any member of our species who has walked the earth in at least the last 100,000 years. That includes all hunter-gatherers and other "primitive" people studied by anthropologists.

It's important to understand this: when we say that hunter-gatherers inform us about our past, it's not because we think they are relics of that past. Countless studies have shown that for all intents and purposes, they are biologically and psychologically just like us. The differences are cultural and ecological, not evolutionary or genetic. Their special interest derives from the fact that—for circumstantial reasons—they persisted in living as we all once did. So in studying them, we can watch our bodies and minds operate in our ancestral environments. In a way, they are our avatars to the past.

Notice that I put "environments" in the plural. We can't talk about *an* ancestral environment; there were many in time and space in those 100,000 years. But the range of those environments does not include what has happened since the invention of agriculture. When we consider the change since the emergence of what we like to call civilization, the contrast is greater, and when we consider the last few centuries, greater still. As we'll see, these changes were not all for the better, and one of the worst was what happened to relations between the sexes. For most of the history of our species, women were in a stronger position than they have been since—stronger, at least, than in almost all subsequent cultures until the last few decades in some postindustrial states. Even these latter-day safe havens for women have a long way to go to achieve gender equality. Hunter-gatherers, too, were far from perfect in this regard, but they were more egalitarian than we have been for thousands of years.

If I have succeeded so far in this book, you will already be convinced that the arrangements we humans have made for male and female sexuality and power are scarcely the only way to go. In fact, the opposite is true: the variety in nature is tremendous. But overall, males are nowhere near as necessary as females; as we've seen, females were the first sex, and any type of sex involves conflict as well as cooperation. In some species males predominate, in others females, and in others the tasks of reproduction are shared about equally. There are species where males have been eliminated, but the same cannot be said of females. Maleness and femaleness rest on evolutionary and biological foundations, but they are not immutable, and there are many lessons to be learned about them from observing the natural world, if we keep our minds open.

If I succeed in the rest of my argument, I hope by the end to have convinced you that what we think of as male supremacy throughout past history was actually a rather late development in human evolution and will ultimately turn out to have been a temporary social adaptation. It was maladaptive in the deeper, more durable past, and it is maladaptive now, going forward. But first we must retrace the steps by which our last common ancestor (LCA) with bonobos and chimpanzees turned into the fully modern humans that we were by 100,000 years ago, and think about some of the biological, especially sexual, novelties that this transformation gave us.

The LCA—let's call her Laca—was an ape that lived six to ten million years ago in Africa. We don't know exactly what Laca was like, but she was one among many ape species that had by then been successful for millions of years. She was at home either in the trees or on the ground, climbing or walking on all fours with grasping hands and feet, and had a brain that was ape-sized or smaller. Something made upright walking an advantage for Laca; viable theories include being able to carry things, seeing farther, displaying newly evolving breasts or (for her partner) genitals, and reducing the heat absorbed from the sun. Modern apes and even monkeys do upright walking at times for some of these purposes. We know it had little to do with making stone tools, since they didn't appear until perhaps two million years ago, and the same is true for big brains.

My favorite hypothesis is that it was good for carrying babies. Great ape mothers walk three-leggedly after their infants are born, using one hand to support the child. Monkeys hardly ever have to do this, because even their newborns cling well. A monkey mother can unceremoniously start walking with a baby in her lap and the baby will reflexively cling to her with all four limbs. Great ape mothers have to provide support for the first few weeks, and this would have turned into infant carrying as upright posture evolved. It would also have allowed for a more helpless infant, born earlier because of a pelvis made for bipedal walking.

In any case, upright walking must have been a good thing, since several different species had evolved it by at least five million years ago. *Ardipithecus ramidus*—"Ardi" for short—is our most likely ancestor from that era. Ardi was about four feet tall, with long arms, a pelvis that proves she walked upright, and a still ape-sized brain, but with feet partly adapted for grasping. In other words, she was just what you'd like to see in a missing link. But perhaps Ardi's most interesting feature was that she was not significantly smaller than her mate. Not only were the male and female about the same size—you can sex fossils by the angle between the bones that come down from the pubis, among other ways—but his canine teeth were also no larger than hers.

As we know (ever since Darwin), species with intense male competition for females are more likely to have males substantially larger than females. These tend to be species (like the elephant seal and red deer) in which dominant males get most of the sex after a tournament-like contest. They may have harems or simply monopolize females in a larger group. Among monkeys and apes, the males in such species tend to have large, sharp canines. Large body size differences between males and females—"sexual dimorphism"—occur in baboons, orangutans, and gorillas. Gibbons, which tend toward monogamy or even polyandry, have less of a sex difference than humans.

Bonobos and chimps have much less of a sex difference than gorillas but a bit more than us. Males in both species, though, have very large testicles compared with gorillas', probably because gorillas have harems, while bonobo and chimp females mate with multiple males. Either way, males compete. Gorillas strive to get mates into their harems and don't have to worry too much after that, so *they* are huge, but their testicles aren't. Chimps and bonobos compete much more for access to eggs, since fertile females blithely recruit the sperm of multiple males, which then vie with one another in a huge microscopic swimming match that often, in the end, comes down to sheer numbers.

So mating anatomy can take strange turns. Testicles don't fossilize, but sex differences in body and, especially, canine size can point to mating patterns. Michael Plavcan has studied this in detail; in 2012, he concluded, "Although strong dimorphism is consistently associated with polygyny and intense male-male competition, a lack of dimorphism is not associated with any single mating system." Despite these doubts, those studying Ardi argue for low levels of male-male competition—perhaps even monogamy, with substantial paternal care.

Either way, our next likely ancestor, probably descended from Ardi and common in Africa around a million years later, may have had males much larger than females. That was the famous "Lucy" species, *Australopithecus afarensis,* more fully upright than Ardi but brainwise (yet again) no more than an ape. Controversy continues about whether Lucy was substantially smaller than her mate. Ditto for the first two species of *our* genus, *Homo*—the evidence is confusing. But the later species of *Homo,* including Neanderthals, had a size difference no larger than ours. Canine size differences mattered less and less as we evolved, because our ancestors were now making stone tools and weapons. If they wanted to scare and tear, they did not need canines.

Chimps and bonobos teach us that big differences in mating habits and gender roles can arise in only a million years, and voles suggest that one genetic mutation can make the difference between a species with couples and fatherhood and a closely related species with neither. This should give us pause when we talk about fossils spanning six or more million years. In that time frame, mating systems might come and go and female fortunes rise and fall with them.

Since the beginning of the transition from hunting and gathering to other forms of social and economic organization, there has been less genetic change, and none that affects our basic biology. When we talk about physical gender in fully modern humans, we know what we mean: a species with males on average 7 percent taller and 15 percent heavier than females, with little difference in canine size; men with more visible facial and body hair, broader shoulders, and greater muscle mass; women with wider hips, more fat surrounding them, and visible breasts (unlike apes, who have only nipples, except when nursing). Also, girls mature and show their adult anatomy sooner than boys do.

Reproductively, human females have two more key adaptations: first, many years of life after cycling stops and, second, well-concealed ovulation. Primates generally have a less distinct estrus than other mammals, but some (chimps and bonobos included) have sexual swelling and clear behavioral signs of receptivity; we have nothing nearly as obvious. It used to be said that human females had evolved continuous sexual receptivity, but this phrase was evidently coined by hopeful male scientists; what they probably meant was that women are randomly receptive across the days of the monthly cycle.

Even this, however, is not quite true, since women on average are less likely to have sex while they are menstruating, even in the modern American middle class. Many cultures discourage it, and there are some where menstrual bleeding is difficult to conceal completely or where limitations on the activities of menstruating women make those days culturally obvious. Since menstruation signals infertile days, we might think of it as a kind of anti-estrus. Also, recent and growing evidence—reviewed in 2011 by Martie Haselton and Kelly

Gildersleeve—does suggest that women unconsciously signal ovulation and that men are affected by the difference.

To take a rude example: studies show that lap dancers earn more money in tips at midcycle—something about them is making the customers put more bills in their garters—but not if they are on birth control pills, which dampen hormonal changes. Men show more jealousy on their partners' likely fertile days, and they find unknown women's clothing sexier and their bodily scents more pleasing during those strangers' fertile days. In a 2014 study, Jessica Tracy and Alec Beall found that the "red dress effect," according to which women wear red or pink more often when they are ovulating, applies strongly in winter but not in summer, when they have the option to draw attention by wearing less clothing. These and many other recent findings show that the behavioral effects of hormonal cycles have not been abolished in our species. But they are less dramatic than ape estrus; menses aside and compared to apes, women are more likely to have sex throughout the cycle.

Some background: Hunter-gatherers, also called foragers, have dramatically lower population density and social stratification than other kinds of cultures. They are also much less likely to have males living near their male relatives throughout life. This is of great potential importance to women's status, since women have more kin at hand and more choice about where and with whom to live. Married couples may reside in groups of the husband's kin, the wife's kin, or both, and such wealth as there is in hunter-gatherer groups—for example, primary access to water holes and valuable tools, ornaments, and musical instruments—may be inherited through either parent. Male direct care of infants and children is greater in hunter-gatherers than in the other cultural types.

Marriage is close to universal in human cultures, but polyandry (one woman, several men) is officially recognized in just a few (the traditional estimate is around 5 percent), while polygyny (one man, several women) is commonly allowed (82 percent). (We'll take up some dramatic cases, including some fundamentalist Mormon sects, in chapter 6.) Despite this, only some men have more than one wife at a time in any culture, so monogamy has been the predominant human marriage form through space and time. This means one man and one woman *officially* together at a time—social monogamy. It does not necessarily mean complete fidelity of either; nor does it preclude several marital bonds in a lifetime. This serial monogamy is partly due to mortality; for most of human history, "till death do us part" meant an average of fifteen or twenty years. But it is also because of divorce, allowed in most cultures and common in many. Pragmatic polyandry (optimal number of fathers more than one) occurs informally in many cultures that do not officially allow it. Father-infant proximity is greatest in cultures allowing polyandry, lowest in those with general polygyny.

Katherine Starkweather and Raymond Hames showed in a 2012 study that most anthropologists have underestimated the frequency of polyandry cross-culturally. It is true that "classical" polyandry—a substantial proportion of marriages being between one woman and two or more men, usually brothers, to prevent the breakup of family farms—is characteristic only of complex societies in Tibet, Nepal, and the Marquesas Islands. But "nonclassical" polyandry is found in every part of the world. In these marriages, more than one man is officially recognized by the community as having legitimate sexual access to the same woman, with or without cohabitation, and—contrary to the conventional wisdom that this has to be bad for men—the husbands take care of each other's children if one of them dies.

It occurs in situations where women are scarce (because of excess males or polygyny) or those in which there is extreme dependence on male contributions of food (as in Arctic hunters). In the former cases, men may decide they are better off sharing a wife and having *some* chance at reproduction rather than none; in the latter, men may realize that they need insurance for their children to survive. Either way, a woman gets to have more than one father contributing to *her* kids' survival both now and in the future, relying not just on their alliance with each other but on their confusion about paternity. She also gets the fun of having multiple mates competing and trying to please her. Although this occurs in only a small fraction of cultures (compared to polygyny, which is formally allowed in the great majority), the cultures where it *is* allowed are often hunter-gatherers. This means that it was probably an option for our ancestors way back in the deep human past.

Some cultures, mainly in the Amazon basin, turn paternity confusion into a virtue, through the custom of *partible paternity*—the belief that a child can and should have multiple biological fathers. These cultures do not recognize singular biological paternity but say that a pregnant woman *should* have sex with more than one man so that each can contribute to the baby's development in the womb. While this can't happen biologically, there is evidence from some of these cultures that children with more than one "father" are more likely to survive. Some successful men may compensate by having extramarital affairs. But Sarah Hrdy's idea of "the optimal number of fathers" certainly looks as if it applies here.

Hunter-gatherers allow polygyny but don't have very much of it. Only among Australian aborigines did more than a third of men have two or more wives. The average for other groups is 7 to 10 percent. Polygynous men in the Australian cultures had young girls or even infants betrothed to them, and as the men aged they wound up with multiple adult wives. Accordingly, men married very late, and most young men were left to their frustrations. Why, anthropologists continue to wonder, did those old men get all the women and girls? It's not clear even now how they got away with it.

But the rest of the world's hunter-gatherers had much lower levels of polygyny, compatible with young women marrying around the time of menarche—at around age sixteen on average, much later than in modern societies. A few years' difference in age at marriage together with the known lower survival rate for males could explain a small amount of polygyny. The unmarried state in these cultures was highly undesirable and almost unknown except transiently and, for women, in old age. But we can reasonably ask, How can there have been egalitarian gender relations if polygyny was allowed?

For one thing, if polyandry was also an option, then it would not seem so unequal even if polygyny was more common. What we would have would be basically a pair-bonding species with departures in both directions. But part of the answer is how the polygyny comes about. We think of it as mainly due to male coercion, and we will encounter societies where this is indeed the case, but in hunter-gatherers men rarely have that kind of power. Hard as it is for us to believe, becoming a co-wife is often an extension of female choice: a woman may be better off as the second wife of a good hunter than as the first of a bad one. But since meat is shared beyond the family in any case, a better explanation may be the good genes hypothesis.

Women, in this view, are attracted to men who have qualities they would like in their future sons—as in one of the explanations for the peacock's tail. They (unconsciously) seek those qualities in the men they mate with—ideally, perhaps, through monogamy with a favored male but sometimes by becoming his second wife or by having an affair with him while married to a lesser male. Why "ideally, perhaps"? Well, another ideal is for women to seek one kind of man—the reliable, androgynous, "good father type"—when they want to marry and a different kind—the erratic, masculine, "bad boy type"—when they just want sex. The best of both worlds might be to marry one and dally with the other. Ongoing studies, even in socially monogamous cultures, show that women pursue all of these strategies. For example, when asked to "choose" a husband in experiments, women prefer androgynous-looking men, but when asked to think about casual sex, they pick much more masculine types. Moreover, when they are ovulating, they lean to the more masculine even more strongly. This research has its critics, but in 2013 Kelly Gildersleeve and her colleagues answered them persuasively, reviewing "a large number of studies providing evidence that women's sexual motivations and mate preferences shift systematically across the ovulatory cycle."

Much about gender relations among foragers was discussed in a pioneering book that grew out of a conference, *Man the Hunter*, edited by Richard Lee and Irven DeVore, which helped kick-start hunter-gatherer studies. But other answers came in a book that followed some years later: *Woman the Gatherer*, edited by Frances Dahlberg. This book pointed out the misnomer, since *Man the Hunter* itself had argued that women bring home most of the bacon—or, rather, the tubers, nuts, berries, and leafy greens. In a number of hunting-and-gathering economies, two-thirds to three-fourths of the calories came from plant foods collected by women, with only a minority supplied by men in the form of animal flesh.

*Woman the Gatherer* showed, too, that archaeologists had been biased in favor of hunting. The stone tools found with human fossils starting about two million years ago were mainly used in hunting and almost certainly made by men. But the tools made and used by women—digging sticks and slings for carrying food or babies—would not have fossilized, although they would have been as vital to survival and reproduction as stone tools.

Chimps in the wild make and use a wide array of tools in many places in Africa: termite-fishing probes fashioned from twigs with their branches deliberately stripped off, leaves crumpled into sponges to soak up water from hard-to-get-at spots, larger sticks to dig in mud, logs and stones to crack nuts, and others; of all these, only nut-cracking stones would be preserved from the distant past, and they have been found in what is likely an ancient chimp habitat. Finally, females use a broader spectrum of tools in both wild and captive chimps, and the same is true of captive bonobos, although they don't use or make tools in the wild. This evidence suggests that Laca (our common ancestor) probably already used or even perhaps made tools of perishable materials, and females may have done more of this than males.

More importantly, the scientists who contributed to *Woman the Gatherer*—Frances Dahlberg, Adrienne Zihlman, Agnes Estioko-Griffin, and others—were interested in highlighting something else that hunter-gatherer studies showed: because of small group size, low population density, and their crucial, highly reliable contribution to subsistence—today we know that it is on average about half, but that is enough—women's voices were always heard. Almost all decisions—the ones we would call economic, judicial, and political, as well as the interpersonal—took place in conversations around the fire at night. And, depending in both sexes on age, reputation, personality, volubility, and eloquence, women as well as men would have a voice. Furthermore, while men often had to be quiet not only on solitary but even on group hunts, women's gathering expeditions were typically peppered by conversations, and these chats built and strengthened alliances, a bit like grooming in monkeys and apes or sex between female bonobos. Women could work through problems without men and present a common front later at the fire.

Some theories suggest that men and women among hunter-gatherers should be much more different and more unequal than they are. In a 2012 study, Frank Marlowe and Colette Berbesque used data from the Hadza, an actively hunting-and-gathering group, to estimate the operational sex ratio—number of males per female capable of reproducing *today*. Given that women are fertile so little of the time in these groups—they are usually pregnant or breast-feeding—there should be more intense male competition, with a larger resulting sex difference. In theory, men should be much bigger than women, something like in gorillas.

Why aren't they? In human females, two powerful physiological factors mute male competition. The first is (mostly) concealed ovulation. This leaves men confused about women's fertile time and enables them to have a lot of sex outside it. Even dominant males wouldn't be able to monopolize women for the whole time they *look* like they might be fertile. The second is menopause. Contrary to common belief, there were quite a few older people in hunter-gatherer societies. Infant and childhood mortality was very high, giving rise to very low life expectancy at birth; but if you lived to age fifteen, you had about an even chance of seeing sixty, and the last birth, on average, was around age forty. Menopause raises the ratio because we have to count fertile men at the later ages when we can't count women. But there are other special things about human mating and reproduction besides unique physiology.

Monogamy, however flawed, is the main mating pattern in hunter-gatherers, and that makes us largely a pair-bonding species, although with great adaptive flexibility. More important, perhaps, is the huge investment humans make in offspring. Hanna Kokko and her colleagues, summarizing fifteen years of research, asked in 2012, "Should a paternally caring male desert his young to try and breed again?" It turns out that "mating investment brings meager returns." In other words, it doesn't pay for hunter-gatherer men to stray so much that they put their existing children at risk.

In fact, Marlowe himself showed that among the Hadza, men do experience just such a conflict, spending less time with their children when there is an eligible young woman around. However, pregnant and lactating women depend critically on their husbands' contribution to their food supply. Fathers also contribute a great deal to their wives and children by hunting—despite its unpredictability—among the Aché, Hiwi, Martu, and other hunter-gatherers.

Kristen Hawkes, another well-respected Hadza researcher, has argued that men hunt only to show off, gain status, and improve future mating opportunities. She has long held that grandmothers matter overwhelmingly in provisioning young children, while fathers matter little. I think this is an unnecessary forced choice. Hawkes is probably right to think that menopause evolved, prolonging women's lives well beyond reproduction, in part to supply grandmothers. And as anthropologist Ruth Mace argued in 2013, it also reduces competition between generations of women for men.

But even among the Hadza, fathers and stepfathers complement grandmothers in caring for the young, while among the !Kung it is impossible to specify which relatives the flow of calories to children is mainly coming from, because so many contribute.

The important point, first made by anthropologist Jane Lancaster over three decades ago, is that among higher primates, only humans have such extensive provisioning of the young, before and after weaning, by individuals other than the mother; the result is earlier weaning and greater survival afterward. Fathers, grandmothers, and other caregivers also provision the mother herself when she is pregnant or lactating. Sarah Hrdy's 2009 book, *Mothers and Others,* showed decisively that we are what biologists call cooperative breeders, and our great success compared to other apes is due largely to others helping mothers. This analysis has been further confirmed in the past few years by Karen Kramer, Nancy Howell, and other scientists.

But we need to look directly at the lives of women in hunter-gatherer societies. The one I know best is that of the !Kung San, or Ju/'hoansi, of northwestern Botswana and nearby Namibia. (They are commonly called Bushmen, which is sometimes considered derogatory.) My late wife, Marjorie Shostak, and I spent a total of two years there in 1969–71 and 1975. We were two among many anthropologists who have visited and studied these patient, resilient, and good-humored people from 1950 through the present, although they are now no longer hunting and gathering. I have followed this research closely, and in 2005—almost a decade after Marjorie's untimely death—I briefly visited our old friends in Botswana with two of our grown children, a moving and eye-opening experience.

Marjorie studied women through life-history interviews, on our field trips together and on a third she made in 1989. She wrote a classic book called *Nisa: The Life and Words of a !Kung Woman* and a follow-up, *Return to Nisa,* published after her death. My own work on infancy involved me with mothers and fathers, and of course we both absorbed much about gender roles from living among the

!Kung. I have already said that there is no single model for our evolutionary past; there are many. The !Kung provide one. They represent hunter-gatherers in some important ways, but not in all.

We were quickly initiated on the day we arrived, in August 1969. A group of women sat in a circle in a village, talking in animated tones when a physical fight broke out between two of them—surprising to us, since the people were known for being peaceable. We later found out that they had been discussing an accusation of adultery in a young couple with a stormy relationship; since there were as yet no children, the husband and wife might divorce. As usual with marital crises in this tightly knit community, this one was everybody's business.

But the fight was not about them. It had been triggered by a woman mentioning another episode of alleged adultery many years earlier. The two (now middle-aged) rivals in that old dispute were present in the group that day, and the knockdown, roll-in-the-sand wrestling match was between them. Other women soon pulled them apart.

This event revealed much about the people's lives. First, there was no privacy to speak of. They lived in bands fluctuating from fifteen to forty or so, depending on seasonal food and water as well as on relationships. Bands split when talking could not resolve a conflict, but they might reunite in a few months. Band membership was through kinship, although there were no strict rules, so two random members might not be related at all; you might be, say, my daughter's husband's sister's husband's mother.

After learning enough of the language, we moved into a grass hut in a village for several months, until it was clear we were more welcome at a distance. But those months were tremendously instructive. Many if not most nights, people collected around a fire in front of one of the huts and talked, sometimes for hours. Women had infants asleep on their hips or nursing; small children often slept across a parent's lap. There might be three or five or eight adults speaking; participation was optional. Voices were active, sometimes argumentative, sometimes lyrical, and they were usually in both male and female registers.

They might be talking about a sighting of migratory antelope tracks by a man or woman that day, and whether they should move the village to follow them. Or about how much water had been left standing in a certain temporary pool after a light rain. Or recollecting memories of an episode of conflict over meat sharing in the past. Or about a hyena heard prowling near the village on more than one recent night. Or someone's illness and whether they should get up a trance dance to try to cure her. Or a young woman's readiness for marriage and who a likely lucky boy might be.

Probably the predominant subjects of these fireside chats were interpersonal issues: a couple not getting along, a teenager being lazy, a good hunter growing a little too full of himself, a woman suspected of flirting with someone else's husband. Gossip? Sure. But this was not just behind people's backs, or not for long. It was open airing of difficulties that, if not solved, could affect everyone. And people expressed their feelings; they sought advice, they helped each other. After falling asleep to enough of these conversations, I came to wonder if the hunter-gatherer era had been one interminable encounter group. Certainly, these exchanges were one of the vital adaptive advantages of the evolution of speech. As my first anthropology teacher, the linguist Dorothy Hammond, used to say, "What would we talk about, sitting around the fire at night, if we didn't have language?"

Were women's voices listened to as much as men's? Richard Lee, one of the greatest of !Kung ethnographers, estimated that men did about two-thirds of the talking in these group conversations, but that is far from a monopoly. Women were not shy about airing their opinions, and if they had arguments supporting those opinions, they could lay them out in full. These conversations were matters of adaptation, deep emotion, even of life and death. How stupid would it be to not listen to all informed opinions?

Did some men have a gender ideology uncomplimentary to women? Yes. Once a visiting colleague and I were interviewing a group of men about their knowledge of animal behavior, on which my friend was an expert. The subject of women's intelligence came up, and one man ranked it below that of a lizard. Others found this funny. But it was just a tasteless jab in an endless battle of the sexes; women had unflattering things to say about men, too. In the battle of day-to-day life, the sexes were more or less joined in confronting natural dangers and human tragedies and foibles alike.

Then, too, women produced about 70 percent of the food and did 90 percent of the child care—the working mother was no modern invention. Actions spoke louder than words. Women's hands more than earned them the right to have their voices heard, and they were—in mixed groups around the fire at night but also during the days when they were away from men. As in other hunter-gatherers, men's hunts were often carried out in silence, but women's gathering expeditions, in addition to supplying most of the food, were an endless opportunity for talk—small talk, serious talk, exchanging ideas and facts, giving and getting advice, expressing feelings, creating and sealing alliances. Men, whatever their prejudices might be, could be only so dominant in the face of those contributions, those alliances.

Yet men do have a certain appeal. Megan Biesele, the leading scholar of !Kung folklore (and a lifelong proponent of their rights), titled one of her books *Women Like Meat.* Men do almost all the hunting in this culture, and so meat is one of the reasons women like men. Another is that the main religious and healing rite, the trance dance, was traditionally male-centric, because—although women have a separate healing dance—in this ritual only men were said to have healing powers, and they would go into trances to activate those powers.

However, the ritual depends as much on women as on men, since women sit in a circle and sing in eerie, beautiful, yodel-like voices and clap in difficult-to-master syncopated rhythms so that the men can dance in a methodical, plodding, marchlike style made up of little jumps in a larger circle around the women, in order to enter trances. Those capable of trances—up to half the men—eventually fall down, at which point their souls are believed likely to leave their bodies. Other men, often in trance themselves, rub their colleagues' unconscious, prostrate bodies to revive them into a more active altered state. Standing, they then stop in succession at each person sitting (whether part of the inner circle of singing women or a spectator in an outer one), lay on hands, tremble while moaning in a low vibrating voice, and then shriek at the top of their lungs, saying the ritual phrase *kow-hi-didi*—at which point particles causing or likely to cause illness are drawn out of the person's body and pass painfully up through the healer's to be shot out from the nape of his neck, back up into the spirit world they came from.

These two admired roles for men became entwined when the excitement of a significant kill being brought back into camp led women to sing and dance, a spontaneous outpouring of enthusiasm that could lead to a formal all-night trance dance. In our time in a !Kung village, we saw that men were obligated to be very modest, silent, or self-deprecating about their hunting success, insisting that an antelope was small and thin when all could see it was large and fat. Credit (and the chance to distribute meat) went to the owner of the arrows, who could be a woman. Yet despite all this, everyone knew what man or men were responsible for a given kill and, over time, which men were more and less successful. The plant foods collected by women—nuts, tubers, berries, and greens—were the reliable staple majority of the diet, but the very unpredictability of major game hunting was exciting.

For women, there was often something faintly sexual about their enthusiasm, and that appeared to be true of trance dancing as well. While muted and mostly concealed, there was a kind of harmless flirtation at times between some of the women and some of the men in a trance. As an apprentice who on two or three occasions succeeded in altering my state of consciousness somewhat during a dance, I felt the admiration (as well as amusement) of some women as they acknowledged me the next morning. However, nothing would cause an apprentice to be dropped from training as fast as being perceived to be taking advantage of this situation or doing anything faintly suggestive while in a trance. So the balance was a delicate one.

Women had their own, separate trance ritual, the drum dance, also healing. Some women had strong reputations as healers and were sought after by women and men alike. Also, women hunted small game—birds, tortoises, lizards, and more—that they came upon during gathering expeditions. In childhood, these successes were a source of great pride. I have a photo of a seven-year-old with a broad smile on her face holding her arm straight up in the air, a bird she has killed dangling from her fingers. And in Marjorie's book, Nisa describes coming upon a kudu foal with a group of children. She grabbed at it but it slipped away, and they chased it. "I ran so fast that they all dropped behind and then I was alone, chasing it, running as fast as I could. Then I picked it up by the legs and carried it back on my shoulders. I was breathing very hard, 'Whew . . . whew . . . whew!' . . . When I came to where the rest of them were, my older cousin said, 'My cousin, my little cousin . . . she killed a kudu!'"

Like boys, girls had dramatic initiation rites around puberty. Wherever the girl might be, on noticing a trace of blood she had to sit silently on the ground and wait. If she was alone, this could be dangerous, since lions, hyenas, or wild dogs could get wind of her. But the women in her village would soon notice her absence, guess its cause, and track her easily. They would carry her to a place near the village, quickly build a small grass hut, and set her inside it. A unique dance would begin, all the women taking part. As day turned into night and the next day, the dance became increasingly raucous and bawdy. Women wore only a small leather pubic apron, which tossed around as they hopped and swayed. They flipped it up to flash their genitals, provoking hilarity, but meanwhile they sang, danced, and clapped continuously. No men could approach. The menstruating girl was required to sit alone in the seclusion hut, never making a sound but getting the message: an uninhibited celebration of womanhood.

As shown by Lorna Marshall, Polly Wiessner, and other anthropologists over many years, !Kung equality was due to the leveling effect of giving. There were formal rules for meat distribution, as well as rules for gifting ornaments, clothing, weapons, and other possessions, which made their way across social networks. What went around came around. It was not that people were selfless but that cultural rules and the open intimacy of social life made possessiveness difficult. Gossip and ostracism are strong deterrents. Differences in "wealth" could exist, with some people having twice as much as others, but the kind of wealth differentials we routinely accept could not have been attained in this culture, even in the most selfish person's fondest dreams. And the general egalitarian way supported gender equality.

Women, as in any culture, bore the burdens of motherhood, but not alone. They held the ideal, quite unusual across cultures, that women should give birth alone—a tribute to their independence and courage—but this ideal was not attained until later births for most mothers, and first births were almost always assisted by other, more experienced women. Demographer Nancy Howell, whose work laid the foundation for many of our studies, published a book with a new analysis of original data in 2010. She showed that families with two or more children could not supply their own caloric needs, but other members of the band closed the deficit by sharing. Families with children were never left to their own devices. Later in life, women, especially, could reverse the flow of food.

Our first field trip lasted twenty months; when we left, people were continuing to argue about whether or not that young couple should divorce, and there had been many other conflicts in between. During our two field trips, we became aware of four women who made suicide attempts. Two ate arrow poison, very unlikely to be a successful method; one wandered off into the bush for two or three days, until people searched for her and found her; and the fourth rubbed arrow poison into a cut, which would have killed her if the poison had not been old. Like some suicide attempts in all cultures, these may have been calls for help. Each of the four was unhappy about the way her husband was treating her. Yet any attempt has some chance of success.

As for homicide, Richard Lee documented twenty-two cases over several decades in this small population. Men were the only perpetrators and most of the victims, and the cause was usually alleged adultery or a vendetta, although sometimes it was a perceived unfairness in meat distribution. How common was adultery? Genetic studies have estimated that 2 or 3 percent of people did not have the biological fathers they thought they had. This is low for hunter-gatherers, but it is not insignificant; few husbands would like to hear that one in fifty times their wives had sex it was not with them. And Marjorie's intimate interviews of women showed that in emotional terms, their preoccupation with extramarital affairs could be absorbing and romantic. Not all women had lovers, but some did.

Nisa, the woman who became the focus of the book, had a life with men that was not representative but also was not extreme. She was around fifty years old in 1970 and had grown up in a very traditional setting. She was married by her parents' arrangement twice before menarche but was allowed to reject these men. She also refused an offer to become a man's second wife, although she loved him. A third arranged marriage was successful, and she learned to love this husband, Tashay, although both went on to have extramarital lovers. He tried to take a second wife, but Nisa drove the woman away.

During her first pregnancy, Tashay became very jealous and questioned the paternity, but he accepted the baby. She had four children with him, two of whom did not survive early childhood. Her son died of an illness later in childhood. One daughter, who survived to be married, died in the course of a physical fight with her husband, who purportedly had not intended to hurt her seriously. Despite infidelities on both sides, Nisa was grief-stricken when Tashay died. She went on to have other relationships, including a bad marriage to a Bantu man who mistreated her and left her but became possessive (as sometimes happens in all cultures) when she tried to move on. She finally remarried and stayed married into late life, though she continued to have occasional lovers. She said, "A woman has to want her husband and her lover equally."

Nisa had more infidelities, more violence, and more tragedy in her life than other women, but she also had autonomy and agency. She said, "Women are strong; women are important. Zhun/twa men say that women are the chiefs, the rich ones, the wise ones. Because women possess something very important, something that enables men to live: their genitals." In Marjorie's view, "!Kung women themselves refer to, and do not seem to reject, male dominance. The fact that this bias exists is important and should not be minimized—but it should also not be exaggerated. . . . All in all, !Kung women maintain a status that is higher than that in many agricultural and industrial societies around the world. They exercise a striking degree of autonomy and of influence over their own and their children's lives."

When I returned with our grown children in 2005, Nisa was alive and welcoming but, in her mid-eighties, poorly sighted and very frail. Yet in a group, she took charge of the conversation, pointing her finger in the air and commanding attention with the fluency of her words and the modulated rise and fall of her still compelling voice. Some of those listening had been children in my original study. One woman, ten years old in 1970, was now a grandmother and had a baby of her own younger than her granddaughter. Her mother and her mother's mother were alive and well—five generations of women in one family.

On the eve of our departure, there was a feast. I bought one of their own cows for them to slaughter, attracting visitors from other villages. A trance dance began, but none of the men went into a trance. Two women did—one the lovely grandmother we had known as a charming ten-year-old. She danced and entered a trance with her child on her hip, collapsed as hands reached out to protect her and the baby, and was revived into a state in which she could lay her hands on people sitting around the fire, tremble, shriek, and by doing this, according to !Kung tradition, heal. I was happy to be ministered to; I felt strangely privileged and, in a way, protected.

Four decades ago, early in what would become feminist anthropology, Michelle Rosaldo and Louise Lamphere published a book called *Woman, Culture, and Society.* In it they said that male dominance is a cultural universal. This seemed a disappointing way to begin a new era, but as they knew and documented, there are differences in just how dominant males may be. It was clear then and remains clear now that hunter-gatherers on average have less gender asymmetry in power than do other kinds of societies. !Kung men, we have seen, try to throw their weight around, figuratively and literally, but women have sound resources of resistance and evasion, including fighting back, rejecting husbands for themselves or their daughters, rebuffing co-wives, giving or withholding sex, having extramarital liaisons, speaking up in group discussions, forming alliances among themselves through talk, cooperation, and gift exchange, and in general taking advantage of the fact that they supply a large proportion of the food.

Except for the last point, which is variable, all nomadic hunter-gatherer cultures share these areas of relative strength for women. Diets vary from one-fourth to three-fourths animal flesh (including fish), with the average being around half and half, and the contributions of the men and women vary accordingly. The lowest plant food levels are found in mounted hunters and in Arctic groups like the Eskimo, which are not possible models for most of the history of our species—no ancient hunters had horses, and we did not penetrate the coldest climates until late in human evolution. But there are also high levels of animal flesh (including meat, fowl, reptiles and amphibians, and fish) in the diets of some warm-climate hunter-gatherers. Yet even where male hunting predominates, small group size, low population density, mutual dependence, and a blending of private and public life restrain male dominance.

The division of labor has never been complete. !Kung men, at a minimum, gathered plant foods for themselves while out hunting, especially if they were not successful on the hunt. Women routinely killed small prey like lizards and tortoises and reported on game or tracks they had seen. Hunter-gatherers living in coastal habitats have extensive shellfish collection by women and children. Among the Martu of Australia, women occasionally kill large game such as kangaroos. Forest hunter-gatherers across central Africa drive prey into large nets, a group effort that routinely includes women. And among the Agta of the highland Philippines, women routinely hunt successfully and bring in about a third of the meat. This example is unique, but in most hunter-gatherer cultures a few women have been avid and excellent hunters. Some groups had a formal status for men who assumed female roles, not just in gathering but in dress, marriage, and other ways.

So despite the fairly reliable division of labor by sex, boundaries have been somewhat fluid and often crossed. This appears to have been the case for a very long time. In a fascinating study called "What's a Mother to Do?" archaeologists Steven Kuhn and Mary Stiner summarized evidence that Neanderthals and the modern humans who replaced them differed in their dietary options, suggesting that Neanderthals were more dependent on large game.

Since Neanderthal females and children (like males) had very robust skeletons, women and children probably joined in the hunt, perhaps driving game toward the hunters. More recent discoveries, however (for example by Amanda Henry, Alison Brooks, and Dolores Piperno), have shown that Neanderthals ground, cooked, and ate barley and other grains; they may even have eaten quite a bit of barley porridge. Women would almost certainly have been involved in this processing, and they may also have made clothing.

Our species, however, shows an even broader dietary range, including more small game, fish, shellfish, and plant foods. This would have led to a division of labor by sex resembling that in recent hunter-gatherers, with women concentrating on plant foods and shellfish. Kuhn and Stiner trace this back at least fifty thousand years, and growing evidence in Africa reaching many thousands of years before that suggests a deep origin of both the dietary breadth and the division of labor.

Hunting of large game is not incompatible with women's roles as mothers; the Agta are proof of that. But their case is unusual. Hunter-gatherers have more involved fathers than other cultures, and the most caring fathers on record are the Aka hunter-gatherers of the central African forest. Even they, however, spend much less time with infants and children than mothers do—aside from mothers' already massive biological investment in pregnancy, birth, and milk. So the need to reproduce meant that women would not commonly hunt large game. Their work produced more reliable returns and helped guarantee their influence and autonomy. Men's more erratic contribution of large kills no doubt generated excitement—as we saw, Kristen Hawkes proposed that its main purpose was showing off—but it did not lead to strong male dominance in any recent hunter-gatherer setting.

The same can't be said of the phases of history that followed. In some parts of the world, at the culmination of the hunting-gathering era, populations became denser, settlements grew more elaborate, and men began to assemble in all-male groups. These groups differed markedly from the nomadic, "classical" foragers that had dominated the planet for almost a hundred millennia. Public and private life could for the first time be separated, and women could be edged into purely domestic roles. Men vied with one another ever more intensely for power, because power now really mattered. The world was no longer egalitarian, either among men or between men and women. Most dangerously, it was no longer possible to resolve conflicts by moving away from your enemies. You had too much at stake in what you and your friends and kin had already built. And not long after that, when you were planting and harvesting, you owned the land in a new way, and you were not going to walk away from it. In fact, you were going to defend it with your life.

*Chapter 6*

✦

# Cultivating Dominance

So violence was not uncommon in hunter-gatherers, but as long as population densities were low and people sparsely distributed and nomadic across landscapes, it did not often lead to war and certainly not to an ideology of constant preparation for war. Men could try to exclude women from influential decisions and assert dominance over them through brute strength, but there were narrow limits to how far this could go. As we will see, this began to change in the last few thousand years of the hunting-and-gathering age, as settlements grew and intensified, but a watershed was coming.

Archaeologist Patricia Lambert, reviewing all the voluminous relevant evidence, showed that the native peoples of North America had violent fights when they were simple hunters and gatherers, and perhaps some group conflict, but warfare emerged with much greater intensity as populations settled down and grew, and especially after agriculture began to exhaust the land's resources. This began in some regions and spilled over to others through conquest; ultimately, the ripples spread throughout regions like the Eastern Woodlands and the Pacific Northwest. The same pattern repeated itself in every part of the continent, with the most war-torn period being the last millennium *before* Europeans arrived.

Brian Ferguson, another archaeologist, showed in 2013 that a very similar process had occurred much earlier in most parts of Europe and the northern half of the Near East: the transition to agriculture, population growth, pressure on resources, organized violence, and war. It is easy to see how a great intensification of warfare provides opportunities for men to dominate women. But the transformation associated with agriculture was not just an intensification of war; it was a complete change in the human way of life.

We can see an instance of that momentous shift in the foothills of the Zagros Mountains in Iran, where archaeologists have unearthed one of the earliest transitions from hunting and gathering to agriculture. In a 2013 study, Simone Riehl and her colleagues showed that by 11,700 years ago, the wild ancestors of barley, wheat, lentils, and grass peas were being gathered in abundance. Large, stable groups began to become farmers. They went on hunting wild boar, gazelles, cows, goats, hares, birds, lizards, and turtles and also feasted on fish, mussels, and freshwater crabs. But the abundance of wild grains they ate increased, so they built bins to store seeds and used grindstones to make flour. Turning seeds into bread or meal was almost certainly women's work, as it has always been in grain-*growing* cultures throughout the world. This was when the women weren't collecting mussels and crabs or gathering wild plant foods. And all of it was while they were nursing babies and tending kids—that is, growing the future.

Around 9,800 years ago, their grains lost the ability to propagate their own seeds—a trustworthy signature of farming, since this means people had to have done the propagating. At this basic level of agriculture, women may have been in charge of sowing, tending, and harvesting gardens; it could easily have been women who invented cultivation. They or their men also tamed the goats they had once hunted; it's easy to imagine a woman adopting a kid whose mother had been killed. For two thousand years, there had been gradually improving blades and other stone and bone tools, as well as stone vessels, and there were clay figurines of animals and people that could have been art, toys, or gods.

Across the Middle East at that time, there were at least five other transformations, in what is now Iraq, Turkey, Syria, Israel, and Palestine, including Jericho; the region was a cradle of farming, but it makes no sense to try to pinpoint the origins, because the genes from wheat to goats show independent transitions even in these places, separated by mere hundreds of miles. Beyond them, Cyprus and Greece were only a little behind. Ideas may have migrated, but the tamed plants and animals did not. This was cultural evolution in parallel. It occurred independently in places throughout the world, wherever a well-watered soil enabled wild plant management, population growth, and the passage from following herds to tending and breeding them. It was also an evolutionary coup for the plants and animals, which reaped great reproductive benefits by cozying up to our ancestors.

This shift, eventually worldwide, began independently in many places: China cultivated millet on the Yellow River, rice on the Yangtze; New Guinea, yams, taro, and bananas; West Africa, gourds and sorghum; Aztec Mexico, maize and squash; the Incan Andes, potato and quinoa; and the Amazon basin, peanuts, manioc, and chili. None of these civilizations had to get the idea from the others. Anthropologist Marvin Harris liked to call the New World the "Second Earth," because it independently repeated the Old World experiment, going from foraging to farming and on to cities and empire. Women's work and, no doubt, their creativity were involved in each transition, yet the consequences for them were profound, pervasive, and none too happy.

But what brought the change about? In 2012 archaeologist Melinda Zeder, building on half a century of research, updated what is called the Broad Spectrum Revolution. This was the worldwide intensification of hunting and gathering that began by about twelve thousand years ago, corresponding to the oldest layer of the Zagros site. Population growth both caused and resulted from the changes; the world had filled up with hunter-gatherers who no longer readily found empty places to go. Global warming after the last ice age played a role. Certain locales with rich, varied resources now allowed more diversified hunting and gathering, based on different abundances around the year. People settled in; they built more permanent homes and communal structures, formed alliances with other groups in the wider area, and had both the need and the power to defend what they had built against intruders. Inequality, political maneuvers, coalition building, and formal armed conflict enhanced men's power.

This new kind of hunter-gatherer culture was seen in real time in the Northwest Coast Indians, like the Tlingit and Kwakiutl. Anthropologist Susan Walter recently summarized what we know about these societies. They harvested abundant salmon, shellfish, and other marine resources; stored food; and had many other resources, enabling a dense population unknown to most hunter-gatherers. This plenty led to social classes, inherited rank and wealth, sizable permanent houses, ceremonies called potlatches with dazzling and even wasteful displays and giveaways of wealth, multiple wives for rich men, raiding, warfare, slavery for captives taken in war, and stunning, ambitious art like totem poles (sometimes with slaves buried under them).

Even in plural marriages, high-ranking women were privileged, doing preferred ceremonial work and crafts, and leaving the drudgery to junior wives and slaves. Women processed food and managed long-term food stores. They smoked salmon—the most important preserved food—and this task required more skill than fishing. Marriages sealed alliances, and wives served as ambassadors to their home communities. Women contributed greatly to subsistence and helped mount the lavish potlatches and feasts—but they did it on behalf of their husbands, the chiefs, who had many wives and slaves. Large polygamous families were favored and led to more married children and more alliances.

Many cultures that sowed the first seeds in a hunter-gatherer world must have resembled these Native American societies. A spectacular site from an early example is Göbekli Tepe, in southern Turkey, which dates to around eleven thousand years ago. Its centerpiece is a group of rings of monumental stone pillars carved with symbolic figures—the world's first known temple—that people made pilgrimages to from many miles around. It was contemporary with the earliest farming cultures, but its builders continued to depend entirely on hunting and gathering. Yet it had already begun to achieve the organization of large numbers of people that would characterize the farmers to come.

The societies responsible for these structures would soon go through a transition like the one in Zagros. Then they would become agricultural chiefdoms, like those in Hawaii, Africa, and the American Southeast—strictly tiered worlds with privileges of inherited rank and wealth, public politics, warfare, captives, slaves, formal religions, ornate ceremonies, substantial dwellings, multiple wives for men who mattered, and death or a subservient role for others. Yes, hunter-gatherers in some conditions had the population and settled life that led to organized inequality, between the sexes as well as among the social classes. But for most of our time on earth, hunter-gatherers were nomadic bands where men may have tried to dominate but women had a voice.

Strangely, the shift to agriculture in many parts of the world worsened health as it increased human numbers. In a 2011 study, Amanda Mummert and her colleagues, including the leading paleopathologist George Armelagos, who died in 2014, summarized a number of new excavations made since Armelagos and Mark Cohen first put forth the case for this counterintuitive claim a quarter century before. Of the newer studies, fourteen showed a decline in height from before to after farming began, while just five showed the opposite, and in these cases other changes explained the trend. The fourteen can be added to the many older studies that showed declines in height, along with other indicators of poorer health.

Generally, the decline can be attributed to increased malnutrition and infectious disease in childhood. Farming made food abundant but narrowed its variety. Hunter-gatherers took advantage of seasonal and local variation in vegetables and fruits and also ate meat and fish, so they had a broader range of nutrients, while farmers often depended on one or two starchy crops. This also made them more likely to go hungry in a dry year. Infectious disease spreads faster in dense, settled populations—although in general, most traditional societies are found to have roughly similar high mortality. But how did the population grow if mortality wasn't lower?

Women had more births.

Hunter-gatherers have later weaning ages and longer interbirth intervals than farmers, and this supports the idea that agriculture brought more frequent births, affecting women dramatically. Patricia Lambert, in 2009 in a paper called "Health Versus Fitness," proposed a basic Darwinian explanation: the decline in health was a price that people making the transition had to pay for greater reproductive success—the only meaning of fitness in Darwinian terms. Certainly, this first demographic transition allowed our species to fill the world at a higher density than hunting and gathering had. But new discoveries suggest other consequences.

Paleopathologist Vered Eshed and her colleagues, in ongoing research, examined the skeletons of more than two hundred Natufians—hunter-gatherers living in what is now Israel and neighboring areas. They compared them to people living in the same places after the switch to farming. They did not study height, but they did look at life expectancy and disease. Trauma was equally prevalent in hunter-gatherers and farmers, although skull trauma declined, possibly reflecting fewer encounters with large prey or less conflict. But infectious, inflammatory disease rose markedly, because of a denser population and rodents infesting grain stores. Hunter-gatherers, for most of their history, were also protected by their nomadic way of life.

The surprise was this: there was no overall decline in longevity in this sample, yet women's life spans *declined* while men's *increased.* Young women were especially vulnerable, probably because they were starting motherhood earlier and having more births closer together. This asymmetry is one reflection of the biological change in women's lives when hunter-gatherers settled down as farmers. Women were probably working harder, but they also were turned into baby-making machines. They had always been the source of new life—this is as old as life itself—but with agriculture, the always-risky pregnancies and births came faster. It was almost as if Eve had gotten Adam's curse as well as her own: she now tilled the soil by the sweat of her brow to get her bread *and* her sorrows in bringing forth children were multiplied. Women gave more of themselves and died younger even as they were cut out of public life.

To be sure, in many places there were transitional, small-scale cultures, relying on gardening and a few domestic animals, or sometimes even a form of agriculture, that left room for women's influence. Anthropologists reported on similar cultures in modern times. In New Guinea, Margaret Mead described Tchambuli women as unadorned, managing, and industrious, fishing and going to market while men fussed with their hair, put makeup on, and went head-hunting; Mead's account has proved to be exaggerated, but sex roles there were certainly not the same as in the West, whose readers she was addressing. Others have studied the Mosuo (also known as the Na), a peasant ethnic group in China, where women shared power because wealth passed through the female line; they kept their husbands literally at a distance—the men were basically tolerated visitors—and occasionally replaced them. The Minangkabau in Indonesia (and the related Malays of the state of Negeri Sembilan, in Malaysia), also matrilineal, reckoned descent from a queen mother; women had more influence than in many cultures—controlling marriage, for example. And the Ashanti of Ghana were a matrilineal kingdom in which women ran markets and passed their brother's wealth to their children.

In matrilineal societies, roughly a tenth of the world's cultures, women had more autonomy, including weaker restrictions on extramarital sex, since men passed most of their wealth to their sisters' and not their wives' children; you don't have to care much about the paternity of your sister's children, since you are equally related to them whoever their father is. In some of these cultures, the dueling dominations of husband and brother gave a woman more breathing space and more control of her life as she played the two men against each other. But elsewhere she was under one or both their thumbs, and polygamy was allowed to them but not to her. So far, there has never been a *matriarchal* society—one truly run by women, not even where chiefs and kings inherited everything from their mothers.

Among the Malays of Negeri Sembilan, for example, men were considered to have stronger *semangat,* which is the gatekeeper of the self, reducing the likelihood of spirit possession. This contrasts interestingly with the !Kung, where men are considered more likely to be able to leave their bodies during trance, but either way the ideology favors men. In summarizing their situation, Michael Peletz, the leading ethnographer of gender in this society, says that both men and women spend much of their lives in a social universe "deeply suffused with ambivalence, alienation, and tension." This is true of all human societies. He also observes, "Through a complex chain of symbolic associations, men and women alike appear to view menstrual blood as indexing the limited extent to which humans are able to 'rise above' their animal natures" and that "menstrual blood, and women generally, represent to men (however consciously) the precarious foundations on which male ascendancy rests." I would be hard-pressed to find a better expression of the way men feel threatened by women.

The men in this culture claim that they have both less passion and more reason than women do, and women tend to agree with them, although there is substantial variation in individual views. Men seem to feel that the differences between them and women justify a certain level of irresponsibility as husbands and fathers. Significantly, wealthier women are more accepting of male hegemony than poorer women are. Men's behavior toward their wives is often compared to elder brothers' treatment of their sisters, and the husbands tend to fall short of the elder-brother ideal. Among the Minangkabau, especially as described several generations ago, the husband's role was even more tenuous; he might visit his wife only from dusk till dawn, and his children might not recognize him in public. A traditional aphorism about husbands said, "Like the ash on a tree trunk, even a soft wind and it will fly away." With them as with their Malay counterparts, we are still left with what anthropologist Audrey Richards called "the matrilineal puzzle"—as Peletz puts it, "how to trace descent through women yet allocate authority to men," and also how to negotiate the power of a woman's male relatives against the claims of her husband. But married men are also under pressure to satisfy the economic demands of their wives' *female* relatives, one of the sources of women's power in this culture.

Despite centuries of Islam, in Muslim Southeast Asia women were less socially inferior than in neighboring East Asia and even, at the time, in the West. Female gods survived, and "Women predominated in many rituals associated with agriculture, birth, death, and healing, perhaps because their reproductive capacities were seen as giving them regenerative and spiritual powers that men could not match." A kind of gender pluralism prevailed, with acceptance of some forms of unconventional gender behavior, relatively relaxed sexual mores (with certain red lines not to be transgressed), and a sense of women's right to sexual pleasure in marriage. Much of this was due to matrilineal or bilateral reckoning of kinship and descent and the underlying belief systems.

If Minangkabau men were traditionally like ash on a tree trunk, this is even more the case with the Na of southwestern China. In fact, for many Na women, you could say that the ash never settled on the tree trunk in the first place. Here, even in recent decades, the husband was often little more than a furtive nighttime visitor, and he might share that status with many other men. More than for any other, this is, as in the title of Cai Hua's ethnography, "a society without fathers or husbands." Even a review in *American Anthropologist* that accused the book of exaggerating Na uniqueness (citing the famous Nayar of northern India and some Polynesian cultures) also said, "They clearly are an extreme case of privileging visits over marriage, and an extreme case of privileging matrilineal ties over marriage ties."

Tami Blumenfield, Eileen Walsh, and others have noted that the Na tend to have long-term partnerships, one at a time, with the visitors, and that these "walking marriages" (the people's own term) lead to paternal responsibility (including financial contributions and direct care), since a woman usually knows who the father is (and would feel embarrassed if she didn't). Nevertheless, the main household itself is matrilineal, and it is run by a matriarch; by comparison, the role of male sexual partners is fluid and small, based on women's affectionate choices, and subject to change. It is certainly a challenge to the Western idea of marriage. Outside the household, however, Na society is stratified and patriarchal, especially at the top, where even descent becomes patrilineal and noble male legacies are preserved.

So despite these fascinating cases, and however they reckoned legacies, early farming and herding cultures were almost certainly patriarchal, even if some were matrilineal, and "Big Men" would have had central roles in their clans. Among the Enga of highland New Guinea, studied for decades by Polly Wiessner, "Relations between men and women were structured by separation of the sexes that assigned public roles to men and private ones to women."

Patriarchy at first may have been weak, but as populations swelled, they grew ever more stratified and unequal. They had to stand and defend their crops and grazing land. Plowing, livestock, and war put male strength at a premium, and those who could built coalitions to rule women and weaker, poorer men. Many patriarchies eventually built pyramids, but even before that they *were* pyramids—social pyramids, and steep ones, too. They were also systems of predatory expansion, growing at the expense of neighboring peoples. In Mesopotamia, Egypt, China, Mexico, Peru, and Nigeria, similar trends independently appeared at different times. All the trends enhanced men's power.

Thousands of years passed between the invention of farming and the rise of empires; that awaited the emergence of cities housing imperial governments, organized religions, a merchant class, and an army under central control. Urban life was decried from the first for its filth, crowding, loose morals, and corruption of youth, yet it became a worldwide trend, unabated since. Cities beckon according to what people can learn and gain, and how and with whom they can play when they take breaks from striving. And for males, it is often about access to females.

Karl Marx famously claimed that "capital comes into the world soiled with mire from head to toe and oozing blood from every pore." Whether or not this is true of capital, what we call civilization *literally* arose from the mire of flooded fertile soil spattered with the blood of conquered peoples. Men killed men and seized women or enslaved both. Wealthy hereditary aristocracies had standing armies and allied with priestly classes. All were coalitions—conspiracies—of men. The masterpieces of art, architecture, and literature that we associate with civilization are compelling, and the artists who created them are worthy of homage, but they were made in the service of violent male hierarchies, not for the simple sake of beauty.

Slaves built the majestic Sphinx and the towering pyramids of Egypt, so that all who might have thought of freedom could only gaze and wonder. The step pyramids of the Aztecs were altars where scores of thousands of conquered victims had their hearts cut out while alive. Magnificent life-sized terra-cotta soldiers and other works of art were buried with Chinese emperors, but in many cases concubines and other servants were, too, entombed alive with the king's corpse to keep him company. The superb craftsmen who created the terra-cotta army that tours the world's museums were killed by the thousands when they finished it, to keep their skills secret.

Mayan painting and hieroglyphs of the classic period are the products of artistic and literary genius, but they often depict human sacrifices made to celebrate the ascension and power of new rulers. The Parthenon was designed to display the strength and importance of Athens, and it was built largely by slaves, who made up half the population—then and throughout the subsequent period we associate with Athenian democracy, philosophy, and theater. The Colosseum is a masterpiece of architecture, but its purpose was to intimidate Rome's subjects and enemies, and to that end countless gladiatorial contests were held in it, to the delight of the nobility and other spectators—a type of human sacrifice. The exquisite temple complex at Angkor Wat, for centuries sacred to Hindu and then to Buddhist nobles, commoners, and worshippers, is completely covered with bas-reliefs depicting battles and conquests, although the military displays are in some parts of the complex relieved by hundreds of carvings of dancing girls.

It does not detract from the achievement of the artists, craftsmen, architects, and engineers of all these and many more revered ancient works to admit that their purpose and function was not just beauty. The artists weren't in charge; they only worked there. No ancient society could have built these monuments without the use of oppressive force, and no civilization would have invested in them purely for aesthetic reasons. They dazzled all who beheld them, seduced or intimidated foreigners, and, along with religion, control of goods, and sheer physical force, helped keep subordinates—women included—under control.

In all early civilizations, women were subjugated, and their status *declined* over time. Royal women could help their husbands rule and, rarely, rule for a time themselves, but their voices were typically weak, while anti-woman ideologies were strong. Noblemen had many wives, and even for commoners, the family was explicitly the father's little kingdom. In Shang dynasty China, a woman joined her husband's family home and worshipped *his* ancestors. Mesopotamian wives had their property controlled by their husbands and were forbidden from extramarital sex, though their husbands were allowed it. Aztec rulers would send raw cotton to their enemies to insult them, implying that they were only fit for weaving—women's work. Several civilizations allowed gay sex but ridiculed a man who took a woman's role.

Fertility was pivotal, both practically and symbolically. Rich men married as many women as they could, and captives became concubines. Goddesses were worshipped, but male deities dominated in myth and ritual, and heavenly females' privileges were not reflected in any earthly mirror. Women could become priests in some civilizations, but they had to be celibate—unless they were lower-class, in which case they might become temple prostitutes in fertility cults. Marginal women were often sex workers unadorned by sacred purpose. Women could farm, trade, or do other valued work outside the home, and noblewomen had more influence in some civilizations than others, but men ruled in all of them.

We often suppose, quite reasonably, that one factor in women's status is how much they contribute to the economy. Anthropologists have been thinking about this for a long time; Judith Brown's "A Note on the Division of Labor by Sex," published in 1970, was a touchstone. She found that the division was universal and made men value women, but she concluded that "repetitive, interruptible, non-dangerous tasks that do not require extensive excursions are more appropriate for women when the exigencies of child care are taken into account." She gives examples of women's plant and shellfish collecting, which do require excursions and are not so repetitive, yet often permit taking babies and children along. However, for most societies, compared to the hunting-gathering era, women did less autonomous, more repetitive work tied to the home.

A few years later, cross-cultural researchers George Peter Murdock and Caterina Provost did a systematic study that remains valuable. In 185 societies, only fourteen out of fifty technological tasks were almost exclusively male. Women had no such nearly exclusive activities, although cooking and processing plant foods came close. Of course, there is that other vital task in which women overwhelmingly predominate, but it isn't technological: baby care and, less exclusively, child care, which has been the greatest perceived value of women in every traditional culture, however changed that may rightly be in the modern world.

Yet of all the fifty activities, only two—hunting large aquatic animals and smelting ores—were the sole province of men. Gathering plant foods, fuel, and water, spinning, and dairy production were among the tasks that were chiefly female, while males predominated in hunting, lumbering, mining and quarrying, stoneworking, metalworking, and boatbuilding. But we see exceptions even in hunting large animals: among the Agta women in the Philippine highlands, who almost rival men in killing game, and among central African foragers, where women are critical players in net hunts. As for fishing, tending large animals, and house building, women did these tasks in quite a few cultures.

Then there were "swing activities"—twenty tasks that could go either way, like fire-making, hunting small animals, preserving meat or fish, weaving, making clothing, baskets, and pottery, or planting, tending, and harvesting crops. These were rarely shared equally in any given culture, but a particular community could assign them to be mainly done by either sex. So even for tasks that *across* cultures were equally likely to be done by men or women, there was usually only one sex doing it; *within* a given culture, a division of labor held. This means that traditional people felt comfortable when women and men around them were doing different things; it didn't matter if the folks over the next hill were doing the roles in reverse. Yet most cultures depending on farming and gardening got both men and women involved in different ways. This means that the move from hunting and gathering to farming *needed* women as well as men to pitch in, and improvements relied in part on women's inventiveness.

But increased childbearing and shortened life span were not the only reasons agriculture made life worse for women. Even more, it was because of men's power politics—male coalitions out of control. Life as a whole was no longer face-to-face; private and public spheres diverged, and every major aspect of social life—politics, economics, religion, defense—became detached from hearth and home. Males assembled, sparred for positions, and scratched their way up the pyramid as best they could; they barred males who couldn't make the grade—and all women, as well.

There is an old distinction in sociology between *gemeinschaft* and *gesellschaft*—community and society. Hunter-gatherers lived in the ultimate communities: you spent your life with people you knew, and collectively you made decisions that in later cultures would be made by specialists—political, judicial, economic, religious, and so on. As populations grew and hierarchies arose, a separate apparatus usually came into being for each of those categories. Men in these new settings could, really for the first time, freeze women out. Hunter-gatherer men may have wanted to but couldn't; women were present for just about everything, and they had their say.

Then, too, in densely populated settlements, with their political hierarchies, organized violence became far more important and so, accordingly, did male aggression and complex male coalitions. If hunter-gatherers had less organized violence than later subsistence types, it was more because they weren't organized than because they weren't violent; we know that they had individual fights, including homicides. But as horticultural and pastoral societies developed into chiefdoms, warfare became a crucial way to defend and augment resources, and to capture slaves, women, and the means to acquire women—bridewealth to present to a woman's family. For Plains Indians this meant horses; for many African cultures, cattle. But every successful warrior who didn't get killed or maimed was also acquiring something else: reputation.

If hunting was showing off, what was success in war? In a simple gardening culture like the Yanomami, anthropologist Napoleon Chagnon found that men who had killed another man in a raid or battle had more children and grandchildren than men who didn't. Among the Ilongot of the Pacific islands, also gardeners, groups of men mounted head-hunting raids on neighboring groups. Ethnographer Renato Rosaldo wrote eloquently, "When a victim is beheaded, older men discard the weight of age and recover the energy of their youth." Meanwhile, "youths advance from novice status and adorn themselves with red hornbill earrings. . . . To wear such earrings, they say, is to gain the admiration of young women and to be able to answer back when other men taunt. And taunt they do. . . . 'Others will scorn you if you marry without taking a head,'" one young man said.

We know that violence, including war, has often been between men over women, but was that true in deep human history? Neanderthals show the scars of violence in many of their remains, but they were different from us; their robustness, seen literally in their bones, could have been in part an adaptation to conflict. Yet the fossil record of our own kind before agriculture also shows violence.

The record is sparse, and even in violent human societies most people die non-violently, so it is remarkable that we see as much as we do in fossils. Homicide has been part of our lives for at least

27,000 years. At Grimaldi in Italy, in a find from that time, a projectile point was embedded in a child's spine. Czechoslovakian cemeteries of around the same vintage show numerous violent deaths. A Nile Valley man buried 20,000 years ago had stone weapon points in his belly and upper arm. Between 14,000 and 12,000 years ago, the era of *settled* hunting and gathering, there are many more such cases in Egyptian Nubia, and violence was common at sites in Europe. The famous alpine "Iceman" of five thousand years ago, whose well-preserved body has been meticulously studied, has an arrow in his upper back. Research suggests that he was alone in the mountains, had a last meal, was hunted down and shot in the back, and bled to death.

Most of this was still in hunter-gatherer times; there were also homicides in many recent hunter-gatherer societies, including the !Kung, Eskimo, Mbuti, Hadza, and others. It has often been said that hunter-gatherers did not have group-level violence, but this claim is no longer tenable; complex hunter-gatherers like the Northwest Coast Indians certainly had quite a bit. Also, historical studies of "classical" hunter-gatherers suggest that their level of intergroup combat has been understated. Finally, southern African rock paintings, Australian aboriginal clubs and shields, and common spear wounds in skeletons thousands of years old in the American Southwest point to hunter-gatherer group violence.

But a study by Douglas Fry and Patrik Söderberg published in 2013 balanced the picture. They examined twenty-one mobile forager band societies in a worldwide sample. They tallied all cases of homicide reported by ethnographers. Most lethal events were one-on-one, and even when more than one perpetrator was involved in killing a victim, it was usually not between groups. Inter*group* lethal events were uncommon, and most were attributable to one culture, the Tiwi of Australia. But when hunter-gatherers settled in at higher population densities with rich resources this picture changed, and in some of these societies war became routine.

After the transition to agriculture the evidence is abundant, and prehistorians have destroyed the myth that early farming cultures were peaceable. In fact, belief in this myth required blindness to evidence, which archaeologist Lawrence Keeley has called "interpretive pacifications." Many skeletons have embedded spear and arrow points, fractures on the left side of the skull (from clubs and axes in the attacker's right hand), and parry fractures of the lower arms from fending off those blows. Many excavations include graves with weapons and armor; fortifications are everywhere. Among the civilizations we looked at—Mesopotamia, Shang dynasty China, the Aztecs, and so on—power was more centralized and the military more effective. These emerged as states, not tribes or chiefdoms, and their social organization resembles that of the legendary rivals of the Homeric epics, the Bible, the Mahabharata, and other revered classics. Going from there to the wars of modern states is largely a matter of technology. History since hunter-gatherers settled down can be seen as ongoing, expansionist tribal warfare. Nationalism, historian Arnold Toynbee said, is "a sour ferment of the new wine of Democracy in the old bottles of tribalism."

The great literary and religious works of all civilizations give a central place to war. "Sing, O Goddess, the ruinous wrath of Achilles," begins the poet of the *Iliad,* and we soon find out why the superhuman warrior is angry—angry enough to take a long break from the Trojan War and watch his own Greeks go down. His rage results from a special theft: a beautiful concubine named Briseis—captured by Achilles after he killed her husband, father, and three brothers—has been seized by the Greek king to warm his own bed. Achilles, insulted, sulks on the shores before the walls of Troy while his fellow Greeks are slaughtered. The king can do what he wants, but the hissy fit of this single hero could cost the Greeks the war.

And why are they there? Because of an insolent theft of a more important woman: the fabled beauty Helen, the wife of the king's own brother, captured and taken from Greece to Troy by a young man—whom she did not, it seems, strongly resist. This nonetheless causes a war with many thousand deaths, recalled in poetry for almost three millennia.

Achilles's anger is not eased, but it is pushed aside by a greater fury: his friend and lover Patroklos is killed in battle, a death that can be blamed on Achilles's own pride. He cuts the throats of twelve captured Trojan boys on his lover's funeral pyre, and when *his* grief turns to rage, Troy is done for. Little wonder that the philosopher and critic Simone Weil called the *Iliad* "the poem of force." But however much the ancient Greek males loved and made love to men, they collected women. Achilles gets his prize lady back, although he is later killed. The Greeks win, and the king goes home with yet another impressive captured concubine, only to be laid low by his own straying, vengeful wife. The course of true love never did run smooth.

Yet, according to tradition, the face that launched a thousand ships was not Helen's but that of the king's daughter Iphigeneia, whose throat *he* slit in order to fill his fleet's sails with wind. Her mother's later slaying of her father will have something to do with his having traded their child for a salt breeze. But how did that lady become the king's wife in the first place? He killed her first husband, raped her, and forced the marriage. Little wonder she took a lover during the king's Trojan decade; together they did the king in when he came home with *his* special sexual prize—a former Trojan princess who was also a seer. She foretold the bloody sequence, including the queen's later death at her son's hands.

But of all these violent events, the only unusual ones were the killings within the king's family. Slaughtering men to get their wives and daughters was routine in ancient war, and women were one main reason for war.

The great moral texts of the Judeo-Christian tradition send a clear message, shown first in the creation of Eve from Adam's rib.

The patriarchs Abraham and Jacob each have multiple wives and concubines. In one of the odder Bible stories, Abraham's nephew Lot gives his virgin daughters to a crowd that wants to sodomize (in Sodom) some guests thought to be angels, and later those daughters seduce their dad, citing their belief that he is the last man.

Jacob's daughter Dinah is captured by the prince Shechem, who lies with and "defiles" her—but falls in love. The youth's father tries to make a deal with Jacob, but Dinah's enraged brothers play a prodigious trick on the men of the town. They say, Okay, marry Dinah, but you have to be circumcised first. The men agree, and while recovering from this delicate procedure are slaughtered en masse by Jacob's twelve sons, who take captive "all their wealth, and all their little ones, and their wives." Most of Jacob's sons then spawn their own large polygamous broods—the tribes of Israel.

Moses himself has two wives, and he issues clear instructions about captives. Numbers 31 recounts, "And the children of Israel took all the women of Midian captives, and their little ones, and took the spoil of all their cattle, and all their flocks, and all their goods. And they burnt all their cities wherein they dwelt, and all their goodly castles, with fire." Moses (whose senior wife is a priest of Midian's daughter) is angry: "Have ye saved all the women alive?" This kind of munificence had caused a plague in the past—God's wrath or, perhaps, a social disease? "Now therefore kill every male among the little ones, and kill every woman that hath known man by lying with him. But all the women children, that have not known a man by lying with him, keep alive for yourselves." These numbered thirty-two thousand.

Chapters 19 to 21 of the Book of Judges—a book rife with wars—repeat some of these themes. A Levite man has taken a concubine who "played the whore against him," returning to her father's house. He retrieves her with a long visit and much persuading. While staying for the night on the way home, he finds men outside his door who want to have sex with him, but he gives them the concubine instead.

They rape her all night long, and the man finds her dead with her hands gripping the doorstep. Bizarrely, he cuts her in twelve pieces and sends them out to the tribes of Israel.

This leads to civil war: "And this is the thing that ye shall do: Ye shall utterly destroy every male, and every woman that hath lain by man. And they found among the inhabitants of Jabesh-gilead four hundred young virgins, that had known no man by lying with any man; and they brought them unto the camp." Now, though, they have to heal the wounds of war and find wives for the surviving men of Benjamin: "Go and lie in wait in the vineyards; and see, and, behold, if the daughters of Shiloh come out to dance in the dances, then come ye out of the vineyards, and catch you every man his wife." And so they did. The tale concludes, "In those days there was no king in Israel; everyone did that which was right in his own eyes."

And when there *were* kings? David, pacing his palace balcony, catches sight of a bathing beauty: Bathsheba, married to one of his loyal captains. He summons her, has his way, elects to keep doing so, and sends her husband to the most dangerous part of the front, where the unsuspecting cuckold is duly killed. This adulterous, quasi-homicidal union produces the great, wise, and prolific King Solomon. The father had at least seven wives and ten concubines, but the son had seven hundred and three hundred, respectively.

I was raised on these Jewish biblical texts, and when I asked about the morality of all this sex and slaughter, the answer was usually some version of "Things were different then." They certainly were, but the difference—fairly described as much greater violence between men over women *and* much greater abuse of women by men—persisted across the globe for thousands of years. As such it was reflected in the cherished literature and sacred texts of all civilizations. The stories were told in such a way as to justify what was done, and they served as excuses and models for ongoing conflict and abuse.

The Mahabharata, the great Hindu epic of rivalry and war, has had sacred status for many centuries. The hero is Arjuna—a skilled archer, diffident but brave warrior, and morally virtuous leader—who is uniquely worthy to have the Lord Krishna teach him the path to goodness, just on the eve of a great fratricidal battle that he undertakes with reluctance and pain. The result is the Bhagavad Gita, perhaps the most essential Hindu scripture. But in the epic itself, the pure and gloried Arjuna has romantic and sexual adventures along with martial ones; he ends up with four wives and a son by each. Polygyny was certainly common among the nobility of ancient India, and it was explicitly allowed under Hindu religious law. Another important text in sacred Hindu law, the Manu-smriti, provides (among much else) detailed instructions about how the king should avail himself of the pleasures of his harem. For example, "The king may enter the harem at midday. . . . When he has dined, he may divert himself with his wives in the harem; but when he has diverted himself, he must, in due time, again think of the affairs of state." No rest for the weary.

Confucianism praised polygyny and helped spread its acceptance from China to Korea and Japan. Buddhism allows it today as a secular matter in Thailand, Burma, and elsewhere, although Tibet also has polyandry—the only country where it's official and common. In Islam, the Qur'ān permits up to four wives, but the man must love them equally and they must have their own property; many Muslim countries permit polygyny today. Like Abraham and Moses, Muhammad was a highly successful and sometimes unforgiving warrior, and within a few generations his followers had conquered much of the world and were able to avail themselves of countless new mating opportunities. And, of course, the Mormon religion was founded in part upon polygyny; although it has long been forbidden by mainstream Mormons, people claiming to be Mormons (sometimes called Mormon fundamentalists) continue it, with many thousands of them in polygynous marriages in the twenty-first-century United States.

The only major religion to have officially opposed it from the outset was Christianity, which nonetheless tolerated it in important men. Charlemagne had three wives, five concubines, and twenty children; Martin Luther approved bigamy for a ruler; and priests and other clergymen, including some highly placed in religious orders, had access to multiple women throughout history. Nonetheless, as we will see in the next chapter, principled Christian advocacy of monogamy joined parallel Roman attitudes to eventually change the laws and customs of Europe and to some extent the world.

Evolutionary anthropologist Laura Betzig has long been interested in men who collect women. Her classic work *Despotism and Differential Reproduction* became a milestone of Darwinian social science by showing that throughout history authoritarian rulers in every part of the world accumulated scores, sometimes hundreds, of wives and concubines and fathered hundreds of children. We will return to these kinds of rulers and the implications of their prodigious breeding, but for now let's look at polygyny and reproductive success in a much broader range of human societies, based on a 2012 update by Betzig.

Her goal in this research was to consider the mean, range, and variance—a statistical measure of variation—in reproductive success in different kinds of societies for women and men. The range starts with zero and goes up to whatever number has been found as the maximum. The variance is a subtler measure that avoids giving a single case too much weight. She compared three kinds of societies: hunter-gatherers, herders and gardeners, and intensive agriculturalists. In the first two groups, only societies with very good data were included. The third group is historical, based on documents, and she reported only the number of children for the top guy.

For women, the maximum number of children may be anywhere from six to eighteen, and in this hunter-gatherer women resemble herders and gardeners. Even in hunter-gatherers, though, men have more children, as well as a greater range and variance than women do. For herders and gardeners, *all* the measures show a greater sex difference than that found in hunter-gatherers; in other words, although the maximum number of children for women is similar, the maximum number for men is greater. And in the historical cultures practicing intensive agriculture—the Incan, Mesopotamian, Hindu, Aztec, Egyptian, and Chinese cultures—the emperors had hundreds of children each. Other nobles—their associates and officers—also each had many wives.

Polygyny, we have seen, is allowed at some level in most human societies, while the reverse pattern is much less common, at least officially. Consider the chiefdom, a transitional stage between simple gardeners or herders and the grand historical empires. Berys Heuer, an authority on the women of the New Zealand Maori, writes that three of their traditional paramount chiefs had eight, five, and twelve wives respectively, while "lesser known chiefs had equally large numbers of spouses." The English arriving at Jamestown, Virginia, in the early 1600s found that the paramount chief of the Powhatan had more than one hundred wives; he was depicted with many of them in a 1612 English drawing. Kamehameha I, who ruled in Hawaii in the early 1800s, had many wives, as did other traditional Hawaiian chiefs. And the currently reigning king of the Zulu people had six wives and at least twenty-five children as of 2012. Even among the matrilineal Ashanti of Ghana, where a man derived his status from his mother, a chief was expected to have many wives. These cases occur throughout the world, throughout history. Powerful men amass wives in proportion to their power.

Or as evolutionary ecologist Bobbi Low put it in "Women's Lives There, Here, Then, Now," her incisive and sweeping overview of constraints on women, "Humans, like most other species, show an overwhelmingly positive relationship between resource control . . . and male reproductive success. . . . For females, the important relationships are between risk aversion . . . and reproductive success." And: "For humans, as for other mammals, polygyny is the 'default' mating system. . . . When some males can dominate mating opportunities, they will."

We will consider how and why Europe chose monogamy (sort of), but for now we'll contemplate the impact on women's status in many societies in which they knew that rich and dominant men could, on a whim, add them or their daughters to a harem as girls and keep them for life. Even in smaller-scale herding and gardening cultures, polygyny was more likely than in hunter-gatherers, and women's influence suffered. Often, their own reproductive success was also at risk. As anthropologists Monique Borgerhoff Mulder (for the Kipsigis of Kenya), Beverly Strassmann (for the Dogon of Mali), and Daniel Sellen (for the Datoga of Tanzania) showed, having co-wives reduces your number of surviving children in some circumstances, depending on how rich the husband, how many co-wives, and other life events. So, many women lost out in polygynous marriages, even as their men were gaining. But in passing, we might also give a thought to the countless men who, in this system, never found a wife or lost their lives in continual wars.

There are four ways that wars have always been partly about women, as the Bible, the *Iliad,* and other ancient classics reveal. First, men must protect their wives and daughters from being stolen or raped by a raiding party or conquering army and retaliate if they can. Second, men want to seize and rape women left behind by their own fallen enemies. Some feminists have said that rape is not sex, and of course it is also extreme violence, but as cultural historian Camille Paglia writes, "Sex *is* power, and all power is inherently aggressive. Rape is male power fighting female power." Third, if a marriage between two rival groups fails, then the peace made by the marriage ends. Fourth and finally, war can bring the reputation and spoils that enable a man to marry back home.

Horrific things have been done to women in recent wars, but, tragically, this is not new. Historians have largely ignored what happened to women in war in the past, seeing it as a by-product of other goals of war. But in a 2011 study, historian Kathy Gaca redressed this balance for the ancient Western world, examining not just literary or religious texts but other documents and concluding, "The violent subjugation of women and girls through sexual assault and torment has been an integral and important part of Western warfare over the two millennia from the Bronze Age to late antiquity." This means from the time of the Trojan War and the biblical Book of Judges until well into the Christian era. Even when the main goals were to seize cultivated lands and precious mines, "the objective of taking captive girls and women as subaltern wives, concubines, prostitutes, and slaves remains central."

Medieval historian David Wyatt studied a later, more focused time and region: Britain and Ireland between 800 and 1200, including Scandinavians (Vikings) who colonized those islands. In all three cultures, the abduction and subjugation of women, including violent rape, was a nobleman's badge of honor, and even common farmers could have female slaves and concubines. To earn and keep a place in the hierarchy, a man had to "have control over the procreative capabilities of a daughter, a sister, a female kinswoman or servant. . . . Men of all social levels would aspire to female accumulation," abbots and bishops included. So pivotal was this process that the Irish word for "female slave," *cumal,* became a unit of value, equal to three ounces of silver, eight or ten cows, or a certain amount of land. No corresponding standard of value was formally derived from male slaves. A Danish king in a twelfth-century chronicle tells his soldiers to plunder England and "cut the throat (regardless of pity) all of the male sex who might fall into their hands, preserve the females for gratification of their lust."

Wyatt draws a parallel to the Mursi, a herding people in southwest Ethiopia near the Sudan border, whose all-male age sets gathered simply to wage ritual war. These, in turn, resemble the Nuer, pastoralists of the Sudan, a favorite subject of anthropologists; they built an effective organization for predatory expansion at the expense of their similar but weaker Dinka neighbors. Yet these were not true chiefdoms, as in medieval Britain and Ireland, where powerful men controlled their subordinates' sex lives in time of peace but preferred to direct young men's desire outward, through war. In these Western medieval settings, illegitimate children of captured women (and there were many) would grow up to be warriors and wives or concubines. Leaders of the medieval Christian church sometimes preached against these patterns but more often condoned or even joined in them. It was only later that centralized royal power favored nominal monogamy.

But centralization changed only the out-group. In the Crusades, capture and rape of women was standard practice on both sides, despite the alleged religious aims of Christian and Muslim alike. The Europeans took women on the adventure, as wives, servants, and prostitutes, and when the armies of Islam captured them, they would be raped, ransomed, or both. Noblemen on both sides had difficulty reinstating abused wives, so the women often were violently raped abroad and then, if ransomed, lost status at home. Prostitutes and other female servants belonged to whichever men prevailed.

This dark history was not quite a pure quest for male reproduction. Men sometimes raped men to humiliate them and sometimes mutilated or killed women after rape; neither of these tactics enhanced the victors' reproductive success, at least not directly. Fertile, even nubile women were buried with powerful men even in peacetime—a waste of wombs that challenges evolutionary theory. But we no longer have to debate whether all this war and conquest enhanced *some* men's reproductive success. DNA analysis of the Y chromosome by Tatiana Zerjal and her colleagues showed in 2003 that about sixteen million men alive today—including one out of every twelve central Asian males—are genetic descendants of *one man* who lived at the time of the conqueror and polygynist Genghis Khan, whose sons and grandsons had similar habits and power.

Most likely, the man in question was Genghis Khan himself. And, as shown by Laoise Moore and her colleagues in 2006, roughly the same proportion—about 8 percent—of Irish men today have a genetic type traceable to *one man* who lived more than a thousand years ago; he was probably a chieftain in the mold discussed above.

Over the centuries those Y chromosomes passed through the wombs of millions of women, many of whom did not freely consent to the usage; nor did millions of disadvantaged, deprived, bereft, and slaughtered fathers, brothers, husbands, and suitors of those women, whose fruitless genes died with them. Not just in Asia and Ireland but throughout the world, men bear the genes and to some extent the inclinations of those who in a durable past killed men and seized women at will. As ecologist Bobbi Low put it, war is a kind of runaway sexual selection, and we have long dealt with the genetic legacy of that process in males. This is not to say that all men are rapists; very few are. But too many men in the modern world have inclinations that reflect the actions of male ancestors who achieved their success by means of violence and male domination. As I will show you later, the differences between average men and average women in sexuality, like the overwhelming predominance of men in violent crime and antisocial behavior, are likely to be in part a legacy of successful men in this benighted past. But those behaviors are no longer adaptive, and we now consider to what extent these powerful, negative forces have been brought under control.

*Chapter 7*

✦

# Samson's Haircut, Achilles' Heel

It might seem that the status of women is a simple concept: high, low, middling, terrible, worsening, improving. But it turns out that the varied elements that go into that status can be independent in culture and history. Do women contribute a lot to the economy? Are they sharing their husbands with co-wives? Can they take formal leadership roles? Do they have a public say or a vote? How about private influence? Do they help choose their children's husbands and wives? Do they own land?

Consider our own recent history. Women could inherit estates long before they could vote, and they voted long before they began regularly to occupy high elected office or run corporations. For generations they joined in public conversations through writing, but their contributions to the economy varied. Their right to choose their husbands would gradually improve, although it always depended on economic circumstances.

Nevertheless, we have a sense of what women's status means in aggregate. No one would contradict me if I said that women in the West have come a long way but have a long way to go. Nor could

I fairly be taken to task for saying specific things, like "Women have begun to be elected to high office" or "There has been a huge flow of women into the labor force" or "Contraception has given women more control over their lives, but the decline of marriage has left them more often alone with their children" or "Women vote more than men do, so they have increasing influence in U.S. national elections."

Similarly, if I attempt a sweeping view of history at this point in the argument, perhaps I can be forgiven. In simple, mobile hunter-gatherer bands, women had a voice; discussions were face-to-face, decisions day-to-day. Men tried to dominate them, but it wasn't easy. Some men had more than one wife, but the great majority didn't. Hunting made men important and enabled them to show off, but it also encouraged them to invest in their children. War, that universal booster of male status, was possible but not widespread or common. Inequality was limited, whether among men or between men and women.

When hunter-gatherers settled in larger, denser populations, inequalities widened. These cultures might have had nobles, commoners, and slaves, and they went to war to get more of the last. Men became more separate from women and children, but they collected women for sex and reproduction. Women were increasingly the objects of male strife, which was often in the end about controlling wombs. Men had the chance to dominate women, and they took it. Politics became a male game, played in public spaces where men could shame, ridicule, and exclude women.

All these tendencies increased with the rise of farming, again with the rise of chiefdoms, and yet again with the rise of empires. It was not a uniform, linear process in all the places where it happened. Those who study these things in meticulous professional detail say there are problems with generalizations. They are right. But in this case, the devil is not in the details; it's in the overview. And the overview is one of declining power and status for women.

One of the ways we know this is because of a classic study by Martin King Whyte, with assistance from Kristin Moore and Patricia Paul. They approached the subject as anthropologists with respect for culture and its subtleties, recognizing that no one factor is a decisive determinant of how well women do, in life or in power. But in the end, they did come to one firm conclusion: the more complex the society, the lower the status of women. Social complexity itself is not easy to define, but it includes inequality among men; social stratification; separation of political, economic, religious, and military spheres; occupational specialization; and surplus production.

Their sample consisted of preindustrial cultures, so the most complex of them were agrarian states and empires with intensive plow agriculture. The plow itself predicts male dominance, as do herding large animals, hunting large game, and war. Probably the worst examples of males out of control are chiefdoms like those we saw in medieval Britain and Ireland, but these *generally* resembled similar societies at various times and places around the world, before cities, states, and empires.

Thomas Hobbes, in his 1651 book *Leviathan,* claimed that life in a state of nature is "solitary, poor, nasty, brutish, and short." If he was thinking of simple hunter-gatherers, the predominant state of nature we evolved in, he was quite wrong. It was anything but solitary, showing intense social commitment and solidarity. It was poor by his standards (upper-crust seventeenth-century England) but usually adequate and better than the lot of the lower classes of London in his time and long after. It was neither nasty nor brutish but often gratifying and always fully human. And it was about as short, on average, as the lives of Hobbes's poorer fellow countrymen, for the same main reason: uncontrolled infections, especially in infancy and childhood.

On the other hand, if Hobbes was thinking of the lives of his own English people a few centuries earlier, thousands of years beyond the hunter-gatherer state of nature, then for most it was indeed (although never solitary) poor, nasty, brutish and short. The chieftains and their friends worked hard to make life like that for everyone else. But however successfully brutish you were, you wouldn't be on top for long. Hobbes argued that the only way to avoid chaos was to have a single, sovereign, authoritarian state—the "Leviathan"—strong and ruthless enough to suppress discord.

If violence waned with the rise of the state, however, it wasn't without long, hard wars, and it needed a blurring of the distinction between the army and the police. Put differently, the hereditary aristocracy suppressed the rest of the people every day in every way. Anthropologists call this *structural violence,* and it persists in much of the world. Men are overwhelmingly responsible for it. Some ancient empires were less brutish than others, and some rulers survived by pulling back on taxation, forced labor, conscription, seizure of women, and other oppressions. But we are talking about varying degrees. And at any given social level, men oppressed and dominated women.

Also, states exported violence to foreigners on the margins even at times when they eased up on abusing their own citizens. This included slave capture—universal in the ancient world, not abolished in Britain until the 1700s, and persisting illegally *everywhere* today—and deployment of women's reproductive capacities more or less at will. Thomas Jefferson's African-American descendants are just a hint of the genetic legacy of male slave owners. In Hobbes's England, in the decades after *Leviathan* helped establish an illegitimate stability, women were tortured and burned as "witches," and there were an estimated 62,500 prostitutes in London, one in five women and girls, some as young as eight or nine years old. This was survival sex, hunger-driven slavery. Many women died horrid syphilitic deaths.

Of course, an authoritarian ruler must punish treason, including religious opposition. Women found guilty were burned at the stake; men were hung but cut down, castrated, and disemboweled while conscious, then quartered. Many Catholic priests met this fate. Executions were public, and the body parts were sent around the country for display. This was extreme state terrorism, but structural violence was routine under almost all regimes in the last few thousand years. Witch hunts targeting women—75 to 85 percent of the victims, who over the generations numbered scores of thousands—joined rape and forced prostitution as a warning against an independent spirit.

Two examples from very different parts of the time line reveal the structural violence that kept control of lower orders of women and men.

The tribes of ancient Canaan—farmers and herders before, during, and after the biblical Kingdom of Israel—were dominated in turn by many empires. The Egyptians took notes. In the valley east of the Carmel Mountains, thirty-five hundred years ago, the pharaoh laid siege to a large town from May to December one year, after which the local princes crawled on their bellies to kiss his feet; then they and their wives and children carried their own wealth to Egypt. The take included 11,000 tons of wheat, 20,500 sheep, 1,929 cattle, 2,041 horses, 924 chariots, 200 coats of leather mail, and 502 bows. From three towns farther north, the Egyptians picked up 1,796 slaves (plus their children), 235 pounds of unworked gold and silver disks, and a lot of finely crafted furniture, bowls, utensils, clothing, and statues. Many men died in battle, so most of the slaves must have been women.

Those allowed to stay home in Canaan faced confiscatory taxation and brutal reconquest. Another Egyptian note: "The scribe arrives. He surveys the harvest. Attendants are behind him with staffs, Nubians with clubs. One says to him, 'Give grain.' 'There is none.' He is beaten savagely. He is bound, thrown in the well, submerged head down. His wife is bound in his presence. His children are in fetters." But the Canaanites or Hebrews were not special victims. Soldiers were recruited from among them to carry out similar taxation at the other end of the empire, where they did the same to the Nubians.

Fast-forward to nineteenth-century Britain. The gracious and cultured Queen Victoria is on the throne, as she will be for sixty-three years, watching over a vast empire. But consider the streets she looks down on from Buckingham Palace, where 222 offenses are punishable by hanging, including such crimes as damaging Westminster Bridge, begging without a license, picking pockets, stealing from a rabbit warren, being in the company of "Gypsies" for a month, being out at night with a blackened face, and "strong evidence of malice" in a child seven to fourteen years old. Lesser offenders were often reprieved, but more than ten thousand people were hanged in that century. Prostitution was still ubiquitous.

This Victorian structural violence style mirrored the ancient form. Extreme inequalities cannot be maintained without armed police and soldiers, who, on command of the most advantaged, brutalize, imprison, torture, kill, and terrorize the least. *Leviathan* limited violence by making it routine and turning it over to fewer and fewer men. As for the least among women, Daniel Defoe's 1722 *Moll Flanders* and Dickens's 1839 *Oliver Twist* were among the English novels that sympathetically depicted the plight of prostitutes, but both fell well short of revealing the worst about their lives and deaths; the full truth was probably too sad and gruesome for the popular press.

Yet by the time of that "Bloody Code," which punished so many minor offenses by hanging, some cruel and unusual punishments had been abolished. Chattel slavery—the formal ownership of a person—had been abolished in the British Isles in the 1700s and was banned throughout the empire in the 1830s. On the other hand, sex trafficking and sexual slavery continued and, as we will see, is perhaps more prevalent worldwide today than ever before; labor slavery also continues. And every war up to the present has included widespread rape and other sexual violence, often including large-scale sexual slavery.

In World War II, Jewish girls and women were forced to serve as sex slaves by Nazi soldiers, Korean and Chinese girls and women were coerced into serving as "comfort women" for the Japanese army, and Soviet soldiers raped German women in numbers estimated from tens of thousands to over a million, including extreme gang rapes resulting in women's deaths. Stalin declared rapes understandable and forgivable after the long march to Germany. American soldiers are estimated to have raped "only" 11,000, but their army was much smaller. Only a few men (mainly African-Americans) were punished. There were rules against "fraternization," but a U.S. field commander supposedly said, "Copulation without conversation does not constitute fraternization."

Not just by rape but also by fraternization all occupying armies have left children behind. Lower estimates of babies fathered by German troops in occupied Europe are in the hundreds of thousands. Allied troops fathered an estimated 66,000 children in Germany in the decade *after* the war, and larger numbers during it. In the Pacific war and after, including the wars in Korea and Vietnam, American men left an estimated 100,000 Amerasian children behind, most condemned to be ostracized along with their seduced and abandoned mothers.

Given all this, it is hard to believe that the twentieth century was, proportionately, the least violent of all centuries for which we can estimate deaths, but it is true. It is much harder to estimate sexual violence during war, but probably this too has declined; even if it stayed the same, the decline in war would mean a decline in rape. We are talking not about absolute but relative numbers, and the denominator in the equation has grown exponentially, while the numerator has not. This proportionate decline has been shown in every form of violence that has been measured, and Steven Pinker, in his masterful 2011 book *The Better Angels of Our Nature,* brought all the facts together and made the strongest case. We will return to this surprising and welcome trend, but suffice it to say for now that two of the main explanations Pinker offers for it—after the structural violence of *Leviathan*—are democracy and "feminization." That is, the slow and incomplete empowerment of women, and the consequent rise of women's values, has made life much safer for women and men both.

But we can't consider trends in the status of women without thinking about monogamy. When Christians boldly declared that having more than one wife blocks the spiritual communion that is the goal of marriage, it may have been a new philosophy, but it was not new law. Greco-Roman codes had established it much earlier. Both the ancient Greek lawmakers and the fathers of the Christian church were building on hundreds of thousands of years of evolved human nature, which, as we saw, left us a mostly pair-bonding species. Hunter-gatherer marriages were about 90 percent monogamous in principle—although, as in all societies, subject to some infidelity.

Recall that no pair-bonding species has ever been perfectly so; stolen copulations belie the romantic ideals people project on pairing animals, from geese to gibbons. A small but significant minority of hunter-gatherer men had two wives, and either sex might stray in many marriages. But compared to settled hunter-gatherers and the chiefdoms that followed, not to mention ancient civilizations, even high-status hunter-gatherer men had modest ambitions; later cultures' chiefs, kings, and emperors had wives by the score and awarded harems to their friends.

Even royal women often had to be co-wives, and captives could only hope they were desired enough to be concubines rather than other kinds of slaves, who were more brutalized. As for men, those at or near the top did as they pleased until displaced. Even the king of the Greeks, we saw, might have to settle for *this* young beauty rather than *that.* But the men at the base of the social pyramid—a wide base—were spear fodder or otherwise out of luck in the game of evolution roulette.

How this changed in ancient Greece is debated, but it wasn't that women became empowered; two thousand years of European monogamy would go by before *that.* And it wasn't because rich and mighty men gave up their carnal perks. Like Athenian democracy, to which it was only loosely related, Greek monogamy rested on a base of slavery and a stark double standard of sexual freedom. As historian Walter Scheidel put it in 2009, "monogamous" men kept concubines. "They were supposed to draw the line at cohabitation. . . . At the same time, married men's sexual congress with their own slave women or with prostitutes was free of social and legal sanction."

Rome was similar, and recognition of the children of extramarital dalliances was optional. Moreover, ease of divorce underwrote a degree of "effective polygyny"—men couldn't have two wives at a time but "could marry several in a row, thereby raising reproductive inequality overall." We'll consider how this works in our own culture, but in ancient Greece and Rome, men had options. As for the Judeo-Christian tradition, Jews allowed polygyny in the Kingdom of Israel, but in the diaspora they followed local laws. In the Roman Empire, where most Jews then lived, they did as the Romans did, but it was only around a thousand years ago that they officially prohibited polygyny, and that did not apply to all Jews.

Jesus speaks against polygyny in the Gospels, but Paul is silent on it, and Augustine, in the fifth century, explicitly allowed it. Many Christian men of means had multiple wives, and some church leaders had liberal access to women. But formal polygyny was forbidden, and this ban strengthened in modern times. Divorce—serial monogamy—was also against early Christian teaching and even today defies Catholic doctrine. The stance of Christianity grew out of Greco-Roman anti-polygyny, but as Christianity went, so did Europe, and because Europe conquered so much of the world, monogamy is now the norm in most of it. In other words, it was partly historical chance.

Scientists, however, don't like historical chance, so there have been many studies comparing cultures, nations, and species, seeking adaptive reasons for this trend. Some hypotheses: the need for social coherence among males; the social chaos caused by single men; their potential to overthrow the elite who hog the women; the need for fathers to invest more in children; male choice of females who can make high-quality offspring; female choice of monopolizing an average man who invests more over sharing a rich one who won't; male mate guarding to block cuckoldry or infanticide; conflict among co-wives; blackmailing by bystanders when they discover stolen couplings; and, in modern times, a negative impact of polygyny on economic development.

So monogamy is not going begging for explanations. Each has something for and something against it. Mathematical models are elegant and can clarify thinking but are only as good as their assumptions, not to mention the data. Cross-cultural and cross-national correlations are real, but the causal conclusions drawn are rarely proved by the correlations. Is economic development a cause or an effect of monogamy? Did reducing cuckoldry or preventing infanticide come first? Was it male choice or female that started the trend toward pairing off?

We don't know. One or more of these causes began to urge mating systems toward monogamy; after that, growing synergies reinforced it. Nonetheless, it is revealing to consider the deep history of monogamy, which seems ultimately to improve the status of women—but not automatically.

Recall that something resembling monogamy emerged at least twice in our past. First, an ancestral species that was polygynous, polyandrous, or (like bonobos) both had to become mostly pair-bonding. As we saw, some fossil experts put this first change some five million years ago, others not until the dawn of our own species around 200,000 years ago. This transition to pair-bonding at the *species* level is called social monogamy, as distinct from perfect monogamy, which never was. Humankind at the hunter-gatherer stage was mostly monogamous, partly polygynous, and imperfect in all cultures—a strong tendency with great adaptive flexibility.

The second shift is historical, beginning with ancient Greece but picking up speed in the past few centuries: "imposed" monogamous marriage. Almost all human cultures have *marriage*, defined as a stamp of social approval on the union, which legitimizes children; it is something no other species has. There are few possible exceptions—the Na of China (discussed in chapter 6), the Nayar of southern India (a warrior caste with an official symbolic marriage followed by permanent separation and unofficial polyandry with visiting men), and the intriguing feature of partible paternity common in the Amazon basin (chapter 5). But of these, only the Na and the Nayar really challenge the definition, and identifying the father or fathers matters in all.

In any case, most people during the hunter-gatherer phase of our history were involved in pair bonds in a group context—not necessarily permanent or exclusive but socially recognized and meaningful. The first shift made by our remote ancestors probably resembled the evolution of pair bonding in other species. When some of our ancestors, much later, went through the second, it was uniquely human and historical, not genetic or evolutionary. But in between was a time of greatly increased reproductive inequality, the second-class status of co-wifery for many women and the complete reproductive exclusion of countless men. The human male bent toward polygyny got out of hand.

To understand the first transition, we turn again to other species. Two important studies published in the summer of 2013 addressed the question of how monogamy evolved, one for mammals generally and the other for primates in particular.

Zoologists Dieter Lukas and Tim Clutton-Brock looked at more than 2,500 species of mammals. They constructed a mathematical evolutionary tree that allowed them to see where and when something evolved independently. Monogamy *independently* arose sixty-one different times, yielding 229 monogamous species, or 9 percent of mammals. (This contrasts dramatically with 8,000 species of birds, about 90 percent.) Primates, at 29 percent, more than triple the mammal average. Although fathering is far more common in monogamous than other mammals, in over a third of monogamous mammals males don't act like dads. In fact, the only social trait present in almost every case where monogamy arose is that the ancestral females started out dispersed and solitary, probably due to food availability. This meant males could gain more by staying with them than by visiting and then leaving to search for others.

In the other study, Christopher Opie and his colleagues looked at 230 nonhuman *primate* species. Their tree also linked monogamy with female dispersal and fathering, but this time the most consistent background for pairing was infanticide. Because primates lactate for a long time, reducing their fertility, males (as we saw) may kill the infants of other males. In these situations it pays for the male who *did* beget the infant to stay with the mother and baby and protect them. In this account, females living separately followed rather than heralded the rise of monogamy. This is not necessarily a contradiction; since primates have more monogamous species than is typical for mammals generally, the pattern may have evolved by a different process in primates than it did in other mammals.

The controversy continued in 2014, with technical exchanges that are beyond our scope. Lukas and Clutton-Brock believe that in primates, as in other mammals, female dispersal is what leads to monogamy, while Opie's group insists on the difference, citing the greater prevalence of monogamy in primates and pointing out that because the slow development and late weaning of primate infants interferes more with aspiring males' reproductive success, infanticide is a more likely strategy for them.

In any case, this is probably more relevant to us, and not just because we are primates; our mated pairs live in social groups with other couples and individuals. Almost all other mammals that have monogamy have an isolated breeding pair or a pair with nonreproductive helpers (often older offspring). This last is *cooperative* breeding, and we humans have it—siblings and grandmothers hang around and help but don't reproduce. However, we also have *communal* breeding, meaning that several pairs reproduce within a larger group. In mammals, including primates, it's a rare combination, but it makes us humans who we are—the most cooperative primate species by far. It promotes formal alliances between groups through marriage, and it fosters cooperation within them. It takes a proverbial village—or, rather, a nomadic band—to raise a child, including grandmothers, grandfathers, aunts, uncles, older siblings, and cousins, but it also helps to have a man.

After departing from the mostly monogamous hunter-gatherer pattern, and after several millennia of dallying with extremes of polygyny, humans began, first in Greece and Rome, to revert to our starting point. But why didn't the rise of monogamy in Europe make women's status much better? Why was the relative egalitarianism of our hunter-gatherer past still unattainable?

First, as we've seen for the Greco-Roman world and the medieval chiefdoms of Britain and Scandinavia, practical polygyny remained (literally) in force. As long as successful men could own slaves, keep concubines, and seize women in war, the male quest for control of uteruses found many direct outlets. Lesser men could be allowed to marry, but they also could be lured by the bright lights of campfires after battles, beside which they could have their way with captive women—at least for those men who hadn't been killed. Male battle deaths supplied surplus women for surviving males in many settings without the inevitable resentment of hordes of men frustrated by the (unofficial) hoarding of women by men at the top.

Second, when a man could have only one legal wife, suitable

grooms became scarce, and brides' fathers came up with dowries. Instead of gaining bridewealth, which in other systems eased the loss of daughters, they were losing not just a daughter but also a small fortune. As Mildred Dickemann cogently argued, these fathers were probably paying for the future reproductive success of their grandsons; the daughters were a conduit that needed to be placed as high as possible in the social scale. Husbands, for their part, could not legitimize their offspring from other unions, so they depended on the one wife they could get and she bore a big weight on her shoulders. At the same time, the marital bond remained an alliance between groups, as it had been throughout history.

To get an idea of what was at stake, consider the speech Juliet's dad makes when she rejects his chosen groom:

*Graze where you will you shall not house with me:*
*Look to't, think on't, I do not use to jest. . . .*
*An you be mine, I'll give you to my friend;*
*And you be not, hang, beg, starve, die in the streets,*
*For, by my soul, I'll ne'er acknowledge thee,*
*Nor what is mine shall never do thee good.*

We know, as he does not, that she has already married another, his enemy's son. What he would say and do if she revealed that, we can only guess. In the end she has to die for him to get the point.

Thus did countless millions of fathers for more than two monogamous millennia dispose of their daughters on a handshake, making an alliance between two families and two fortunes that no one could put asunder, even the bride herself. There are other Romeo-and-Juliet stories with happier endings; falling in love has been called a "discourse of defiance," a sort of temper tantrum by which two impassioned young people sometimes derailed arranged marriages—but not the ones in which important alliances were at stake, and not if you fell in love with a current enemy.

Women in those monogamous European centuries were as dependent as they had been under polygyny. In ancient Greece, where imposed monogamy began, they got little advantage from it in most places. Sparta was an exception; there, women could be educated and own and manage property—probably because men were so much away at war. In "democratic" Athens, on the other hand, women had no legal personhood. Until marriage, they were under their father's guardianship, and after it, their husband's. They were allowed to enter contracts up to the value of about a hundred pounds of barley so they could engage in petty trade. Male slaves had a path to Athenian citizenship, but no woman, slave or "free," could achieve that status. Aristotle said that wives were bought, and he denied that their household labor had value.

Roman women *were* officially citizens but could not vote, hold office, or serve in the military. Marriage transferred them from their fathers' to their husbands' legal power until Julius Caesar's time, after which they remained in their fathers' power throughout life. Adultery was defined as sex between a married woman and a man not her husband, as in the Ten Commandments of the disdained Jews and Christians. Wife beating was grounds for divorce, but women who became entertainers or prostitutes gave up their right to be protected from abuse.

From the twelfth century on in English common law, a woman who married gave up all her property to her husband. Married Frenchwomen were subject to legal restrictions not officially lifted until 1965. Although Shakespeare's fictional Lord Capulet was mistreating his daughter centuries earlier in Italy, the speech could have been made by a father in Shakespeare's England—or in almost any country in civilized (and very violent) Renaissance Europe.

Two centuries later, in the early 1800s, all the drama of Jane Austen's novels stemmed from the unchanged, dismal facts of life for women. Lizzy, the protagonist of *Pride and Prejudice*, is one of five Bennett daughters who will be destitute unless they find rich husbands; her sardonic but fond father and their relationship are something like the opposite of the situation with Lord Capulet and *his* star-crossed daughter, but his hands are tied. The law says his estate goes to a distant cousin, a self-important nerd who clumsily woos Lizzy without success but with the smug fallback assurance that he will own her legacy. She and her four sisters are at the mercy of unmet men.

Or take the most accomplished and beautiful young woman in *Emma,* Jane Fairfax, pitied by all because her adoptive parents can't leave enough to marry her off; she must become a governess—a celibate slave by the lights of her mean-girls social set. Will an unnamed possible beau save *her*? In the two hundred years since Austen's genius found improbable public expression by edging her hand across a little desk in a busy family room, women have seen more change than in the previous two thousand.

The Enlightenment laid a foundation of *ideas* about rights, and the American and French revolutions set some in law. But American rights were for property-holding white males. The prodigiously gifted Abigail Adams, wife of our second president, wrote him in 1776, when he was just a brash rebel, that she longed to hear independence declared.

> And by the way in the new Code of Laws . . . I desire you would remember the Ladies, and be more generous and favourable to them than your ancestors. Do not put such unlimited power into the hands of the Husbands. Remember all Men would be tyrants if they could. If particular care and attention is not paid to the Ladies we are determined to foment a Rebellion, and will not hold ourselves bound by any Laws in which we have no voice, or Representation.

These were nearly the same words in which the men themselves, three months later, would define their relationship to England's laws. Her "by the way" proved sadly apt: they did not "remember the ladies," so this brilliant woman's influence was in the end through her husband and son, who would also become president. A day when *she* might take the office and be wisely advised by *them* could barely be dreamed of.

Meanwhile, on the other side of the ocean, in 1791 Olympe de Gouges published her *Déclaration des droits de la femme et de la citoyenne*—Declaration of the Rights of Woman and of the [Female] Citizen—which mimicked the *Déclaration des droits de l'homme et du citoyen,* parodying that great decree of the French Revolution point by point. She knew how unlikely it was to extend to her. The preamble cites "the natural inalienable and sacred rights of woman" and decrees, "The sex that is superior in beauty as it is in courage, needed in maternal sufferings . . . recognizes and declares . . . the following rights."

Seventeen articles, as in the "Rights of Man" declaration itself. In place of "Men are born and remain free and equal in rights" she proclaimed, "Woman is born free and remains equal to Man in rights." In the tenth article, she all-too-presciently declared, "Woman has the right to mount the scaffold; so she should have equally the right to mount the rostrum." She would not achieve the latter right, but she did keep the former. She was guillotined two years later, equal, in this way at least, to many out-of-favor male revolutionaries.

But women were making claims that could not remain abstract forever. In 1792 Mary Wollstonecraft had published *A Vindication of the Rights of Woman,* arguing in favor of educating girls. And in 1819 Emma Hart Willard wrote a "Plan for Improving Female Education"; her ideas were implemented two years later in a school in Troy, New York, with municipal funds. Soon high schools for girls opened in Boston and New York City, and some colleges admitted women within a decade. Jane Austen had exposed the plight of dependent women in the early 1800s, but by the 1840s, laws were passed in Britain and the United States shielding a married woman's property from her husband and his creditors.

In 1848, a year of revolt in Europe and the publication of the *Communist Manifesto,* the first women's rights convention was quietly held in Seneca Falls, New York, producing a declaration in favor of equal treatment and voting rights for women. Olympe de Gouges had declared as much two generations earlier, but now a movement began, and it included men.

The next year Elizabeth Blackwell became the first woman to graduate from medical school in the United States, and women were legally allowed to practice. In 1865 the University of Zurich became the first in Europe to admit women undergraduates. In 1869 Elizabeth Cady Stanton and Susan B. Anthony founded the National Woman Suffrage Association, which, half a century later, would win women's right to vote.

Also in 1869, John Stuart Mill, a distinguished British philosopher and member of Parliament, published his essay "The Subjection of Women": "We are continually told that civilization and Christianity have restored to the woman her just rights. Meanwhile the wife is the actual bondservant of her husband; no less so, as far as the legal obligation goes, than slaves." Mill was mocked for introducing a bill that would change the word "man" to "person" in British law; it failed by more than two to one. In the United States, the slaves had been freed and in the previous year granted citizenship by the Fourteenth Amendment. The next year the Fifteenth would give black men the vote, but neither amendment applied to women of any color.

Very slowly, men were won over—as they had to be if they were going to vote to dilute their own power with women's suffrage. Of the 272 men who made up Mill's Parliament, 76 voted for "person" even while the rest of them were laughing. Men *had* to vote in favor of letting women vote. This happened in New Zealand and several U.S. states in the 1890s, and, as a result of steady, brave, nonviolent activism, the Nineteenth Amendment was ratified in 1920, giving all American women the vote—behind their sisters in Finland, Norway, Canada, and Britain.

Meanwhile, education had begun to make its mark. Marie Curie had won *two* Nobel Prizes in science (physics 1903, chemistry 1911), a feat not matched by a man until 1972. Jeannette Rankin became the first woman to serve in Congress, before women had the vote in most states. Julia Morgan was a leading architect who designed buildings for the University of California at Berkeley, among other major projects. At the start of the twentieth century, just fifty years from Elizabeth Blackwell's triumph, there were about seven thousand women doctors in the United States, 5 percent of the physician workforce. By 1920 women had been admitted to more than one hundred law schools; about 2 percent of practicing attorneys were women, and women were almost half of all college students.

In perhaps the most important achievement, Margaret Sanger launched a distinguished career advocating, teaching, and popularizing birth control. The Victorians had debated family planning and ridiculed it: one postcard showed a woman beating the stork off with an umbrella, oblivious to the babe dangling in a sling from the bird's neck. The main method was periodic abstinence, so Sanger and her sister were in uncharted territory when they opened their clinic in Brooklyn in 1916, advocating diaphragms, sponges, and other forms of contraception they had studied in Europe. The basic information they gave out was illegal, so they were arrested. The judge ruled that women did not have "the right to copulate with a feeling of security that there will be no resulting conception." Sanger refused to cut a deal by promising to refrain from repeat offenses, so she spent thirty days in a workhouse. But a later appeal allowed doctors to prescribe birth control, and a new era began.

Giving women this kind of control over their own uteruses undermined male privilege in a way unprecedented in human and prehuman history. The separation of sex from reproduction even a little meant the dawn of a new biology. Now women did not just have the model of Lysistrata, who in Aristophanes' play persuaded the women of Greece to withhold sex until men stopped an ongoing war. They had a new model that did not involve giving up sex, just giving up or, more often, postponing and limiting babies. In evolutionary thinking, neither sex theoretically should want this, but many of both did. However, the impact on women's lives was far greater, and this was only the beginning.

Consider the entry of women into the (paid) labor force. The working woman and working mother are older than our species. But hunter-gatherers, gardeners, herders, and farmers did not have paid labor, just women who worked hard every day. In the Middle Ages and the Renaissance, craft guilds were almost all male, and with the Industrial Revolution few women at first obtained paid factory work; wage labor was in the public sphere, and women belonged at home. This was especially true of married women, whose heavy labor had to do with maintaining households and bearing and raising children.

How could women once again become the influential working mothers they'd been during the hunting-gathering era? Those women had fewer children because of prolonged frequent nursing, in turn made possible by the compatibility of gathering with baby and child care, along with cooperative breeding. The mobility that came with industrialization took nuclear families and, worse, single-mother families out of their kinship context, making work outside the home less accessible than ever. Nevertheless, change occurred.

Unmarried women were in the labor force, but almost all married and stopped working when they wed. This makes *married* women's paid labor a key barometer of women's participation. As economist Raquel Fernández summarized it in her 2013 study, "white married women's labor force participation was at around 2 percent in 1880 and increased very slowly to 1920, averaging 1 percentage point per decade. It grew somewhat more rapidly between 1920 and 1950 (on average 4.9 percentage points per decade), and then took off between 1950 and 1990, increasing on average 12.9 percentage points per decade. Since then, it has stayed relatively constant." Today more than seven in ten married women are in the labor force; unemployment among men exceeded that among women in 2013, as it had for several years.

This doesn't mean that equality has arrived—far from it—but change has been exponential. Consider *all* women's paid labor. Economist Dora Costa noted that in 1900 only 20 percent of women and less than 6 percent of married women worked for pay.

> In the first few decades of the twentieth century, the "factory girl" set the stage for the unmarried "office girl." [She] paved the way for the entry of married women into the labor force in the late 1950s, even though this entry was primarily in dead-end jobs in the clerical sector. [This] paved the way for the rise of the modern career woman, doing work that requires a lengthy period of training and that offers genuine opportunities for promotion.

Change at the professional level has recently been equally dramatic:

> As late as 1970, only 14 percent of all doctoral degrees were awarded to women, only 8 percent of all students enrolled in law schools were women, and only 8 percent of all medical school graduates were women. By the end of the 1990s, women earned 40 percent of all doctoral degrees and represented over 40 percent of all graduates from medical and law schools.

Let's look more closely at what women's work meant. In the 1800s nonwhite women had higher levels of wage labor than whites, immigrants higher than native-born. Many were servants, seamstresses, or sex workers, but the rise of mill towns, beginning in the early part of the century, drew many young women away from farms into mills. By the end of the century, factory and clerical work dominated women's paid labor, although the jobs dead-ended at marriage.

Conditions were horrendous, and women fought to improve them. Some unions, like the International Ladies' Garment Workers Union, were made up mostly of women. Strikes were the weapon in the struggle for more pay, shorter hours—a forty-nine- versus a fifty-three-hour workweek, for instance—and safer working conditions. A massive strike in the fall of 1909 involved twenty thousand blouse makers, mostly young women. A young reporter watched the picket line, where teenage Clara Lemlich, a Yiddish-speaking firebrand, led girls and women.

> [They] began singing Italian and Russian working-class songs as they paced in twos before the factory door. Of a sudden, around the corner came a dozen tough-looking customers, for whom the union label "gorillas" seemed well-chosen.
>
> "Stand fast, girls," called Clara, and then the thugs rushed the line, knocking Clara to her knees, striking at the pickets, opening the way for a group of frightened scabs to slip through the broken line. Fancy ladies from the Allen Street red-light district climbed out of cabs to cheer on the gorillas. There was a confused melee of scratching, screaming girls and fist-swinging men and then a patrol wagon arrived. The thugs ran off as the cops pushed Clara and two other badly beaten girls into the wagon.

The reporter followed other picketers to the union hall, "where one bottle of milk and a loaf of bread were given to strikers with small children in their families. There, for the first time in my comfortably sheltered, upper West Side life, I saw real hunger on the faces of my fellow Americans in the richest city in the world."

The strike lasted from November to February and improved

working conditions. "In the immigrant world, the shirtwaist makers had created indescribable excitement: these were our daughters." They inspired the male-dominated cloak-makers union to strike too.

A turning point came after the Triangle Shirtwaist fire of 1911. A blouse-making factory in New York City went up in flames, killing 146 workers, mostly Italian and Jewish women, in less than twenty minutes. Most burned to death, but a few found another path:

> [A] young man helped a girl to the window sill on the ninth floor. Then he held her out deliberately, away from the building, and let her drop. He held out a second girl the same way and let her drop. He held out a third girl. . . . They were all as unresisting as if he were helping them into a street car instead of into eternity. He saw that a terrible death awaited them in the flames and his was only a terrible chivalry. He brought around another girl to the window. I saw her put her arms around him and kiss him. Then he held her into space—and dropped her. Quick as a flash, he was on the windowsill himself. His coat fluttered upwards—the air filled his trouser legs as he came down. I could see he wore tan shoes.

Here were young women working their lives away in low-wage, dangerous jobs as hard as any done by men, and yet at the moment that those very jobs killed them, they accepted the courtly gestures of a gracious, equally doomed man who offered to help them die.

In World War II, when the male work force was drafted, women took factory jobs not open to them before. "Rosie the Riveter" flooded the workplace with women who now had half the nation's survival on their shoulders. The 1950s saw a return to the prewar status quo, and a population boom to go with it; women were once again thought more crucial in the home than in the world. But on the whole, the twentieth century lumbered in one direction, and as the first baby boomers entered their late teens they made a cultural revolution.

Part of it was a sexual revolution triggered by oral contraception. Part of it was disgust with war, the draft, and nuclear brinksmanship. Part of it was an intense, noble quest for racial equality. Part of it was just rock and roll, with its relentless challenge to tradition and propriety. And part of it, the most far-reaching and important perhaps, was a growing realization that half the human species had been held back long enough.

There are still lessons, perhaps new ones, in some very old stories. Samson was the superhero of the Hebrew Bible and its people, able to do great deeds with his pure male strength. But in exchange for superhuman powers, he had pledged not to cut his hair. While he dallied with the seductive and cunning Delilah, she wheedled his secret out of him. She promptly cut his hair and his heroic career short. Her countrymen captured him, put out his eyes, and imprisoned him, although he did (with God's help) make a suicidal comeback, bringing his Philistine enemies' temple down on their heads.

Achilles, the Greek superhero of the Trojan War, was also uncannily strong, and he was almost invulnerable. His mother, herself the immortal daughter of a sea god, had taken the expedient of dipping her infant son in the River Styx, the stream that separates earth from the underworld, life from death. This made him impervious to harm, with an exception: his mom held him by the heel as she dunked him. He would be killed near the end of the war by the very man who had started it by stealing Helen—a crowning irony for the matchless warrior, who took the adulterer's fatal arrow in his naked, undipped heel.

Both tales deal with the limits of male strength. Both reveal the dependence of even superheroes on women—mothers and lovers. Both teach us that the man—and it has always been a man—who lives by strength and violence alone will die by them, often because of something to do with sex. And both foretell an end to the hubris of unbridled, simplistic, classic masculinity. Physical strength matters little in today's world, and martial prowess is less and less admired as a solution to human problems. For our two superheroes, the end came in their lifetimes. For us as a civilization, it is just ahead; yet unlike with Samson and Achilles, it will not mean a tragic end for men, just a triumph for women. But before we try to characterize the future, we must return to our biological starting point and delve more deeply into the different natures of women and men.

*Chapter 8*

✦

# The Trouble with Men

Change comes because it resonates with the nature of women and men, the impossibility and undesirability of continuing the tragic millennial tale of extreme masculinity, known euphemistically as virtue. Virtue is being redefined, and the old version, closely tied to male political and martial ambition, is being consigned to the trash heap of history. The word comes from the Sanskrit *vira* through the Latin *virtus*, "vir" being also the root of "virile" and "virility." Once, you could not be virtuous without being strong in a primarily manly sense, and that meant treating your enemies in a way we would now deem unvirtuous. For Machiavelli, virtue demanded pretense and treachery and was inseparable from political skill and physical strength. It was also implicitly male.

So part of our story is the feminization of virtue, the lightening of its darker shades, because at their worst they are self-destructive and stupid. It's not that we're free of enemies and the need to outsmart them. It's that the path to victory has changed, and with it the structure of all societies in conflict. We will not win merely by being good, but we will no longer win mainly by scheming, physical bravery, and violence. Before we look at that future, though, we must first see how girls and boys become women and men; then we can turn to leveling the playing field.

If I succeeded in chapter 1, you will agree that the notion of psychosexual neutrality at birth was wrong; very few serious scientists believe it anymore. Some of the behavioral and psychological differences between women and men are the result of hormones acting both in the embryo and at puberty and beyond. But of course there is such a thing as gender socialization. Every culture so far in history has tried to make girls and boys grow up differently, exaggerating the biological differences. As for the latter, we should be celebrating, not minimizing, them or belittling women because of them, and I will show you why. But first, what do we know about how gender socialization works?

In Freud's quaint theory, fear of castration for boys and penis envy for girls were central to gender identity. Anthropologically, there seemed more evidence and logic in Margaret Mead's concept of womb envy; bearing and bringing forth new life is more enviable than having a penis, and in some cultures, like many South American ones, this is formalized in the *couvade:* a man fakes going through childbirth while his wife is really doing it. So cultures differ in gender appraisals. But behavioral sex differences—in aggressiveness, nurturance, toy preference, voluntary sex segregation, and other measures—exist independently of culture and ideology. In fact, as I and many other modern liberal parents have learned the hard way, young children resist efforts to make girls and boys the same.

How *do* they diverge? Differences between the sexes at birth are small, so it is unlikely that they could drive parental behavior. But there is a big difference in the way parents view and treat the two sexes. The first question asked after a birth, the first news, is the one that couldn't be answered for Herculine Barbin: girl or boy? It's as if we need that label to know how to act toward a newborn, and even to know how to *see* it. In monkeys, the first reaction to a birth in many species is that adults inspect and sniff the baby's genitals.

This perception shapes adults' behavior. In experiments, the same baby brings out very different adult responses if it is dressed in pink or blue, and the same audio of a child's mischievous remarks draws amusement and encouragement from adults who think the child is a boy but more negative comments from matched adults who were told it was a girl. There is even a test for parents' gender stereotypes, the Parental Sex Typing of Newborns, or PASTON, Scale. Fathers do more of it than mothers, but both stereotype.

There are also differences in behavior early on. In a study of three-week-old infants and their mothers in middle-class U.S. homes, boys cried more, were more active, looked at their mothers more, and slept and vocalized less than girls, but maternal behavior also differed: mothers of boys held, burped, rocked, stimulated, aroused, stressed, looked at, talked to, and smiled at their baby more than mothers of girls did. Such differences could easily have affected the infants by the age of three weeks; for instance, boys may have slept less because their mothers aroused them more.

Three or four years later, in early childhood, children use developing cognitive skills to label themselves male or female, partly by identifying with same-sex adults, partly through stereotypes. Because young minds tend to split the world, they are less flexible about girls' and boys' roles than they will be later. Confirming older studies on other ethnic groups, May Ling Halim and her colleagues in 2013 found rigid gender typing at age three (as shown by dress-up play among other things) in African-American, Mexican-American, and Dominican-American children. Rigid thinking increased at age four in all three ethnic groups but declined in some ways by age five. Also in 2013, Carol Martin and her colleagues published a study of same-sex playmate preference in 292 children (about two-thirds Mexican-American) in Head Start programs. As in other ethnic groups, same-sex preference was strong, but part of the reason was common attraction to gender-typed activities.

Fortunately, these stereotypes weaken. Clare Conry-Murray and Eliot Turiel, in a 2012 study called "Jimmy's Baby Doll and Jenny's Truck," found that children aged four, six, and eight were able to show some flexibility about gender; they thought that girls and boys should be allowed to choose and this flexibility increased as they got older. Yet self-segregation by sex continues throughout childhood. Since friends and playmates, not just parents and teachers, socialize children, voluntary sex segregation can reinforce stereotypes. Psychologist Eleanor Maccoby thought of same-sex groups as two cultures of childhood that sustain stereotypes even when adult models are neutral.

There are also other influences. In our society, television affirms gender stereotypes. Cartoons, commercials, and other televised materials shape behavior. According to Common Sense Media, in 2011, over a third of babies under age one in the United States watched television at least once a day, over 70 percent by ages two through four. About 40 percent had a TV in their bedroom, and the same percentage lived in a home where the TV was always on. They watched an average of an hour and forty-five minutes a day, not counting video games, online content, tweets, and other media with similar messages, although TV continued to dominate. Repeating the study in 2013, the researchers found a dramatic increase in the use of mobile devices. The amount of time spent using them in all children under age eight had tripled, from five to fifteen minutes a day. The percentage of kids under two who had used a mobile device went from 10 to 38 in two years. Total screen time declined by around twenty minutes, to just under two hours a day, and TV still dominated, with 58 percent of all children watching at least once a day.

Clearly this kind of exposure has the potential to create gender stereotypes. Yet certain sex differences have been seen in so many different cultures for so long—physical aggression and nurturance, for example—that they can have little to do with electronic media. But what if parents the world over were trying to *produce* the same sex differences? In one cross-cultural study, child-training practices were rated from ethnographic descriptions in the databases of the Human Relations Area Files, whose members include nonindustrial cultures chosen to represent the world of traditional societies. Parents in 82 percent of cultures *expected* more nurturance from girls, 18 percent expected no difference, and *none* expected more from boys. Parental expectations for achievement and self-reliance were also consistent but in the opposite direction, with 87 percent expecting more achievement from boys and only 3 percent expecting more from girls. This study did not look at aggressiveness training, but in nurturance, at least, widespread child-rearing practices reinforce a biological tendency.

Though not always. In anthropologist Carol Ember's classic study of Luo children in rural Kenya, boys and girls were assigned different kinds of chores. Girls got water and firewood, made fires, cooked, served food, and tended younger children. But some families had too few girls for these chores and gave them to boys instead. In standard observations, these boys showed less aggression and dominance, more altruism, and less dependence than boys who had not been assigned girls' work, shifting their behavior toward that of girls. In particular, boys who had done a lot of baby tending showed very little aggression.

Since it was only by chance that a family lacked daughters, it is likely that the boys who did girls' work actually changed because of it—that training and experience can make boys more nurturing and less aggressive, whatever the underlying biology. Big increases in fathers' involvement with children in the West in the past few decades and the success of parenting-education programs for fathers point the same way, as do experiments showing that even male monkeys can become more nurturing as they gain experience with infants.

So although biology causes boys to be more aggressive and less nurturing even before puberty, these predispositions are further shaped by experience, training, and expectations. But training and expectations in many cultures widen biological divergences. Anthropologists John and Beatrice Whiting showed many years ago that cultures with frequent combat separate fathers from wives and children—until their sons are old enough to join the separate men's houses, where they begin to train for war.

We've already seen that parents and teachers are not the only cultural shapers of gender. The sexes self-segregate by age three and increase their distance throughout childhood, even as they explicitly *define* gender *less* rigidly. This paradox has yet to be explained, but I'm inclined to think that while kids become too smart to hold on to the extreme stereotypes of early childhood, they continue to want to play with others like themselves—who look and dress like them, prefer the same toys and games, and are growing up to become either women or men. Maccoby highlighted these "two cultures," and as shown by anthropologists Alice Schlegel and Herbert Barry, they decisively diverge by adolescence in most societies. Although children in hunting-and-gathering cultures spend a lot of time in mixed-sex, mixed-age groups, there are some separate games for boys and girls even there.

The availability of same-sex playmates limits options, but with enough children, self-segregation happens. Adults can *make* girls and boys play together, but we have to be pretty controlling to prevent them from drifting apart. Even in cultures with coed schools, children choose separation in free play. In the cross-cultural study by Schlegel and Barry, as girls get older they also tend to hang with adult women, rather than forming teenage groups, but boys tend to form peer groups and keep aloof from men. These patterns make the sexes diverge even more.

So males and females start life with the tendency to be different, yet the sexes always overlap for any measure, and every behavior is influenced by culture. Girls and boys can theoretically be molded to be more feminine, more masculine, more divergent, or more similar. Although no culture has tried to reduce sex differences to zero, except in some parts of the West in recent decades, it is theoretically possible to do this or even, perhaps, to reverse some biological inclinations. But we now know the answer to a famous question of Margaret Mead's: What would happen if boys and girls were raised exactly alike? This is currently being tried, and although it is not an orderly experiment, we have learned enough to say that when they are raised alike, they turn out different.

Many of us who have refused to bring home toy guns have watched our sons and their friends pick up sticks in the backyard and "shoot" each other. Girls who have never been dressed in a feminine way may ask for frilly clothes. Some parents who have given Jenny her toy truck and Jimmy his baby doll have watched in amazement as they swapped their presents. That doesn't mean that these preferences are universal. As a small boy, my wife's nephew wanted a toy tea set; his parents wouldn't get him one, but my wife, Ann, did. He immediately wanted to play with it but, touchingly, he hesitated. "Annie," he asked, "can boys play with tea sets?" She told him they could, and he eagerly did. Incidentally, today he's a typically masculine young man fresh out of college. And if he weren't, that would be fine with us, though it wouldn't be because he used to play with a tea set; that line of reasoning, we now know, is based on a wrongheaded notion of how gender develops.

As we saw in chapter 1, fundamental gender identity and sexual orientation are partly independent of each other and of child training; they are grounded in genetic, neural, and hormonal events that we are coming to understand; this knowledge should make us feel comfortable in our skins and proud of who we are. One of the most impressive discoveries of the last decade in child development research is that when babies of either sex are adopted by lesbian or gay couples—and this has now been studied very extensively and carefully—the main way the resulting children differ from controls raised with a father and a mother is that they turn out to be less homophobic. A 2014 study by Susan Golombok and her colleagues also suggested that gay adoptive fathers are more involved and positive about parenting than are heterosexual controls, but they don't seem to influence their children's gender development. There are serious limits to how much we influence the gender identities or sexual orientations of our children by providing models, environmental influences, or training.

Boys and girls really are different, and so are the men and women they become. It is not, for me, a cliché or a pleasantry to say that I think we are very fortunate as a species to be able to acknowledge that. It is a deep biological and philosophic insight, and although I did not at first accept it—I was a strong cultural determinist in my youth—I am glad to embrace and defend it now. We will see in the next chapter how to make better use of it than we have throughout history—by providing children of both sexes with every opportunity and by focusing on something we really can make a difference in: their sense of equality and fairness. But the difference will endure, and it is not one that favors men.

One of this book's most distinguished predecessors was Ashley Montagu's *The Natural Superiority of Women.* Montagu was a biological anthropologist who contributed to our understanding of adolescent growth and the evolution of childhood, but he was also a steady, helpful voice in the great ideological debates of the twentieth century. He did much to combat the myths of race that scarred that century and to challenge the idea that human beings are hopelessly prone to violence and war. Because he understood biology, he was able to fight those who used it as a weapon against the weak. He lived from 1905 to 1999, nearly spanning the century.

His book on women was perhaps his most exceptional contribution to the conversation about biology and society. He was adamant:

women were not just equal, they were superior, and every apparent piece of evidence to the contrary—the first edition was published in 1953—was the simple result of men's bullying, envy, dissembling, oppression, and abuse. But the book was not mainly about how men kept women down. It was about women's intrinsic biology and why it is just plain better.

Of course, he recognized men's biological advantages: greater size and muscle mass, a higher basal metabolism, the ability to make sperm throughout life, performance in sports that demand speed, muscle, and fierce bursts of energy. But women's advantages were much more impressive; women were longer-lived than men, had lower mortality at all ages, and were more resistant to many diseases, both infectious and chronic. This implies greater fitness and better adaptation. And there were specific achievements as well, even in sports. Because of their fat stores and specialized metabolism, they are superior at long-distance swimming and other endurance challenges. In a dramatic demonstration of this, Diana Nyad, in the summer of 2013, became the first person to swim from Cuba to Florida without a shark cage; she was sixty-four.

Most importantly, women had in their bodies the ability to reproduce—really reproduce, not just donate a micropacket of genes. They could conceive, carry a hugely burdensome pregnancy to term, cope with its massive physiological challenges, give birth with great courage in the face of pain and danger, and produce, for years if need be, a nourishing, disease-suppressing fluid from their bodies that is the ideal food for human infants. And in most cultures they did all this without slowing down very much in the rest of life.

But doesn't this lead to the infamous "Kinder, Küche, Kirche"? This Nazi slogan about women means "Children, kitchen, church"—or, freely translated, barefoot, pregnant, and in the kitchen. No, and for most of human evolution, it never did. Montagu hated and fought this idea, and one of the first things he did in the book was to list women's great achievements, brought up to date in successive editions until 1999. For example, Nobel Prizes in science: Marie Curie, physics 1903, chemistry 1911; Irène Joliot-Curie, chemistry, 1935; Dorothy Crowfoot Hodgkin, chemistry, 1964; Maria Goeppert Mayer, physics, 1963; and in physiology or medicine: Gerty Cori, 1947; Rosalyn Yalow, 1977; Barbara McClintock, 1983; Rita Levi-Montalcini, 1986; Gertrude Elion, 1988; and Christiane Nüsslein-Volhard, 1995.

Today we can add: in physiology or medicine, Linda Buck, 2004; Françoise Barré-Sinoussi, 2008; Carol Greider and Elizabeth Blackburn, 2009; May-Britt Moser, 2014; in chemistry, Ada Yonath, 2009; and in economics, Elinor Ostrom, 2009. Each of these women came of age in a climate that ranged from somewhat hostile to starkly exclusionary and demeaning; few men who have ever won a Nobel Prize had to fight against such odds merely to get the right to study and maybe, just maybe, make a discovery. But most of these women did not only make a discovery; they founded new fields of science.

Montagu also listed nine women who won the Nobel in Literature; now there are four more. And he listed nine who won the Nobel Peace Prize, where we now have sixteen. It is beyond our scope to say what these women accomplished, other than this: theirs were among the most important human achievements since the start of the twentieth century, and they were made by people with one hand tied behind them. Or as someone said about the famous movie dance team, Ginger Rogers did everything Fred Astaire did, but in high heels, going backward.

What Montagu meant was: *We have not yet begun to see what women can do.* "May it not be," he asked, "that women are just about to emerge from a period of subjection during which they were the menials of the masculine world . . . in which the opportunities and encouragements were simply not available?" He went on to quote Oscar Wilde: "Owing to their imperfect education, the only works we have had from women are works of genius." Wilde was a professional exaggerator, but there is more than a grain of truth here. Who but the most inspired, the most impassioned, the most disciplined, the most willing to sacrifice, and the most truly gifted could win on such a steeply tilted playing field?

In the 1950s, Montagu had to contend with lingering "scientific" myths. For example, it was (and is) undeniably true that men's brains are larger than women's. But those who point to this size gap somehow forget to mention that it is relative, not absolute brain size that matters; clearly, bigger creatures have bigger brains. If size were all that mattered, whales would be much smarter than we are. What you do if you are serious is divide brain size by body size before you make your comparison. When you do that, humans have larger brains than whales, and women have larger brains than men. This latter difference is not great, but it is enough to slam the door on male smugness about brains.

We know more about the gendered brain now, and for the most part, the facts don't favor males. For one thing, almost all common brain defects affect boys much more often. This includes autism (4 or 5 boys to 1 girl); intellectual developmental disorder, formerly called mental retardation (2 or 3 to 1); attention-deficit/hyperactivity disorder, or ADHD (3 to 1); and conduct disorder (3 to 1). Psychiatrist and neuroscientist Thomas Insel, head of the National Institute of Mental Health, said in 2005, "It's pretty difficult to find any single factor that's more predictive for some of these disorders than gender." In considering life-course-persistent conduct disorder (leaving out transient teen delinquency), the ratio is 10 or 15 to 1. These few persistently pathological individuals—less than 10 percent of us but doing half the crimes—are almost all male. Women have more emotional disorders, such as depression, anxiety, and post-traumatic stress, but even so, men are three times as likely to commit suicide as women.

None of these is a single-gene disorder, but in those, males again "win" hands down. That's due to X-chromosome deficiency, which leaves a boy's single X free to turn any adverse mutation into disease.

Girls are protected doubly, because they have two X chromosomes and because most of their cells inactivate one of them randomly; for them, normal genes in normal cells counteract most mutations. They would need extreme bad luck—the same mutation on both their X's—to have the disorder, so X-linked conditions are far more likely in males, including congenital cataracts, color blindness, deafness, juvenile glaucoma, hemophilia, hydrocephalus (water on the brain), mitral stenosis (a heart valve defect), nearsightedness, and several neural defects and mental deficiencies, including fragile X syndrome, the second most common genetic mental impairment.

In addition, a remarkable 2012 study proved that the surplus of genetic defects long known to occur with older parents is mainly due to older fathers. This is because the mother's eggs are produced before her own birth, while the sperm contributed by the father divides throughout his life, every division offering a chance for a new mutation. For babies with a mutation neither parent has (that is, one that occurred in the egg, sperm, or embryo), fathers contributed almost 80 percent; and the mutations rise exponentially with the father's age. A thirty-six-year-old father passes on twice as many as a twenty-year-old; a seventy-year-old, *eight* times as many. This means that our history of polygyny and serial monogamy has contributed to many more mutations and defects than we would have had if young men had married women around their own age and stopped reproducing when their wives did.

Meanwhile, the biological resilience of women has led to a current life expectancy at birth of eighty-one for women and seventy-six for men in the United States. Women in their seventies outnumber their male age-mates by around 4 to 3, but by the late nineties it's more than 2 to 1. This means more women are spending their later years without men, and most of them do just fine. By the way, the difference in life expectancy and the difference in health at all ages can make up for at least five years of career time committed to the "mommy track"; no one should be allowed to discriminate against mothers. As Judith Brown showed in a classic study, women in cultures throughout the world have often gained influence after their children grew up, as with international leaders like Margaret Thatcher, Indira Gandhi, and Golda Meir, as well as American leaders like Sandra Day O'Connor, Nancy Pelosi, and Hillary Clinton. This list will greatly expand in the years to come.

Meanwhile, we must give up the illusion that men and women are the same. True, we are similar in many ways, and there is hardly a trait of human emotion or mind where we do not greatly overlap. But we must understand the ways the sexes differ and, especially if you are female, make no mistake about them. Ashley Montagu cataloged many ways in which women are biologically more robust and resilient; the complementary statement is that men must be biologically more frail and vulnerable. But there are darker traits, two especially, that Montagu didn't dwell on, and I do. I will argue, too, that we cannot develop daughters into the roles for which they are destined—let us say for now that this means no more than true equality—unless they fully understand males and how males are different.

The two traits are violence and driven sexuality. Now, the former is sometimes needed, and the latter often is. But as we have already seen in the sad record of history, the excesses of male violence have been literally thousands of times what could ever have been needed, were it not for the fact of *other* males' violence. Men have always been able to justify their violence by reference to that of their enemies. As for male sexuality, it has been, as Camille Paglia pointed out in *Sexual Personae,* one of the great sources of male creativity. But the way it normally (and I use the word advisedly) drives men's actions, the way it becomes detached from relationships and affection, and the way it is often intertwined with violence make it a force that women should respect and fear. I hope at this point you will not take me to mean that they should allow that force to control them; on the contrary, *they* must control *it*, and that starts with wary understanding.

As we saw in chapter 1, physical aggression, however measured, is greater in males than females. By age three in most cultures, boys are rougher and less nurturing than girls. In the United States, Raymond Baillargeon and his colleagues found the difference at seventeen months (aggression in 5 percent of boys versus 1 percent of girls) and saw it at the same level at twenty-nine months in the same children, which is not consistent with a learning hypothesis. This difference persists throughout growth and is greatest in young adulthood, as the frequency of the acts declines but they become more deadly. This is not because males hunted during our evolution. Predation has little to do with aggression, which involves different psychological states, uses different brain circuits, and deploys different actions. Predation in species from big cats to little rats looks like puzzle solving, athleticism, or play more than fighting, and human hunting involves none of the emotion that infuses interpersonal violence. The main use of physical aggression in nature is competition for resources, especially mates. Recall that one explanation for the invention of males is that females gained variation and quality by exporting risky rivalry to a sex incapable of creating new life.

There have been other proposed behavioral or psychological sex differences: in reliance on others for help; sociability; nurturance and affiliation; offering help and seeking companionship; and verbal and mathematical ability. Few have ever held up to scrutiny as biologically grounded. Judging from recent meta-analyses—sophisticated statistical summaries of many studies—most "sex differences" the average person can think of are not even real, much less biologically based. But findings have always been more consistent for aggression, nurturance, and sexuality; hundreds, if not thousands, of studies over more than half a century confirm them. In chapter 1 we saw how hormonal development influences gender identity, even when it is not typically male or female. This is evident in certain specific behaviors as well.

From adolescence, males have much higher levels of testosterone, females much higher levels of estrogen and progesterone—cycling monthly, steeply increasing if pregnancy occurs, and relatively suppressed during lactation, when prolactin and oxytocin are high. In experiments, testosterone increases or enables aggression in humans and many other animals, and male castration reduces it; the same hormone also disrupts normal mothering in females.

However, many species show sex differences in aggression *pre*-pubertally, when sex hormones are at very low levels. Here we have to recall that in humans and other mammals, before or shortly after birth, males have high testosterone levels. These create sex differences in the hypothalamus; artificial early injection of testosterone masculinizes females. In addition, growing evidence shows sex differences in some other parts of the brain. With few exceptions, across primates and other mammals, males are consistently more violent—even in species where females dominate males. As for us, aggression by girls and women does exist, of course, but at much lower levels. In 2013 British psychologist Anne Campbell published an analysis of it, integrating evolutionary and neurobiological perspectives, and showed that it is an evolved characteristic built into the female brain but of a different nature and for different, limited purposes. Human males are more physically aggressive in all cultures at all ages. Socialization plays a role in this, and the outcome results from interactions between culture and biology. But today there is no reason to doubt that brain and hormonal development are foundational.

What about translating violence into war? I have already said, and I am not the first, that all wars are boyish. One of the most unsettling works of modern fiction is William Golding's *Lord of the Flies,* a novel that recounts the descent into extreme bullying and viciousness among a group of schoolboys isolated from adult supervision and forced to survive on a remote island. Could this happen in real life? A classic study called the Robbers Cave Experiment shed light on this question, using twenty-two typical eleven-year-old boys, all from middle-class Protestant families with similar education levels. The summer before sixth grade, the boys were placed in a camp in Oklahoma's Robbers Cave State Park.

In the first week, they were randomly split into two groups. Rivalry was discouraged, and there were many joint activities. Despite this, the groups began to compete—they named themselves Eagles and Rattlers, insulted each other, and were hostile to rivals who entered their "territory." Next, competition *was* encouraged. The groups vied in baseball, tug-of-war, tent pitching, skits, treasure hunts, and cabin inspections; trophies, medals, and four-bladed knives were awarded to winners. By day 3 of this phase, the initial good sportsmanship had been replaced by increased name calling, hurling of insults, and demeaning of the out-group. Soon, abusive stereotypes and negative attitudes toward the other group became crystallized.

In the last stage of the three-week study, the groups were blended again and assigned joint tasks; in one, they were told that vandals had damaged a water tank and they had to work together to fix it so all would have drinking water. Prejudice and conflict declined. In stage 2, when the boys had been asked to name their friends, there had been practically no crossover. But the split had largely healed by the end of stage 3. The boys went home with their biases behind them and friendly feelings restored.

Many experiments with adults in less dramatic controlled conditions show how easy it is to foster bias against arbitrarily created groups. Giving people frustrating experiences or lowering their self-esteem worsens prejudices. Artificially created bias can be reversed when groups are brought together again with appropriate interventions, but in the real world, enemies often stay apart. It is easy to see how the grim process that played out at Robbers Cave in a matter of weeks might operate over generations.

No one has repeated the Robbers Cave study with girls, but a wealth of research has compared the sexes on group identity bias, experimental and real. Women are simply less susceptible to these biases than men. Melissa McDonald and her colleagues reviewed the field in 2012, developing the "male warrior hypothesis." The facts are that men are more xenophobic and ethnocentric than women, more inclined to dehumanize out-groups and use animal words about them, and more willing to make sacrifices to punish them. Across many cultures, men show more social dominance orientation, prefer group-based hierarchy, are more likely to identify with the flag or colors of their group, and are more likely to complete an open-ended statement beginning "I am . . ." with group identity. Men have a lower threshold than women for triggering intergroup conflict and are more likely to order preemptive strikes without provocation in war games. Men, but not women, increase their support for war when primed with an attractive member of the opposite sex. The evidence also supports the "out-group male target hypothesis": male biases and hostilities get stronger when men's attention is drawn to out-group males.

McDonald does not give women a pass on these emotions; she situates them in evolutionary context. Given the human past, women, too, should be wary of out-group males but should relate it to sexual vulnerability, so it is not surprising to find an effect of menstrual cycle hormones. Women are more defensive in their thinking during fertile days of their cycle, avoid situations that might put them at risk for sexual assault, and are more likely to see strange men as sexual threats. More specifically, in one study led by Carlos Navarrete, white women expressed more negative opinions of black men when their fertility risk was high. In fact, the monthly curves for racial bias and conception risk were almost identical. This effect applies not just to the black-white divide but also to other in-group–out-group comparisons, including experimentally fabricated arbitrary groups.

It is increasingly difficult to explain the growing volume of evidence on sex differences in violence, as well as in "groupthink" in conflict situations, without reference to evolutionary theory and biological facts. It is impossible to deny that males show more bias than females. But what was adaptive once is not adaptive now, and if men are not part of the solution, they must be part of the problem.

What about the sex difference in sex? In a short chapter entitled "The Sexual Superiority of the Female," Montagu wrote that while "social conditioning plays a considerable role . . . there is a profound biological difference between the sexes. . . . The male seems to be in a chronic state of sexual irritation. The woman who in a letter to Kinsey described the race of males as 'a herd of prancing leering goats' was not far from the truth." Montagu understood what makes women prodigiously sexual in their own way: the extreme innervation and sensitivity of the clitoris and its short refractory period enable multiple and prolonged orgasms—men are no match for women there—and extend sexual function late in life. But, he believed, women's arousal is not continuous and impersonal; it is framed in relationships and works best when a partner, male or female, makes an investment that comes from caring.

Everything we have learned in the decade and a half since Montagu's last edition confirms these differences. Very few men tell women what they are thinking, and this may be a good thing. Women who "get it" may sympathize (up to a point) or not, but either way they are better off than those who believe men are just like them. Even in post-industrial cultures prominent men often display their intentions by taking trophy wives, using supermodels as "elbow candy," and having illicit sex with much younger women, even in the public eye. But only occasionally do they *say* something revealing.

Henry Kissinger, Nixon's national security adviser, was believed to have more power than the secretary of state when a January 19, 1971, article in the *New York Times* noted that "as a 47-year-old divorcé, he makes society news by squiring such glamour girls as Gloria Steinem in New York, Joanna Barnes and Jill St. John in Hollywood, and Barbara Howar in Washington. Power, he has observed, is the great aphrodisiac." The article depicts him as insecure despite his power; even at his peak he was a short, chubby, plain-looking man. But this did not interfere with his associations with much younger, very gifted, sought-after, beautiful women, then known without irony as "glamour girls."

Bill Ackman, currently in his late forties, is a handsome billionaire hedge fund manager married to a beautiful woman his own age, with three children, and not known for any untoward behavior with women. Perhaps that's why he could be frank in answering his own question in a lecture at the Wharton business school a few years ago. "What motivates people to succeed?" He paused. "Sex. People don't like to admit it but it's the primal driver. Fundamentally, what drives most human behavior is basically foreplay." On CNBC a year or so later, anchor Becky Quick, who had read the quote, offered Ackman a chance to explain himself. He replied, "I was just thinking, you know, well, ultimately we're animals, right? Motivated by basic . . . you know, why are people motivated to succeed? I mean, just think about guys you went to college with."

If that was true of the guys they went to college with, you don't even want to think about the guys in college today. In the summer of 2013, the *New York Times* ran a story with a provocative photo of a reclining woman's bare thigh and the headline "Sex on Campus: She Can Play That Game, Too." Based on months of interviews and authored by Kate Taylor, it was one of many recent writings suggesting that women want casual sex and multiple partners as much as men do.

The article was about the sex scene at an elite university. In a year of impressive work, Taylor had interviewed sixty young women. She wrote about them generally and in a few cases in particular. I read the article to the end and could only guess that whoever wrote the title hadn't. It began with a young woman texting her usual hookup guy to ask what he was doing; she ended up having sex with him. "We don't really like each other in person, sober," she said, adding that they couldn't even sit down and have coffee. She said she was too busy succeeding in school and building a high-powered career to have a relationship and emphasized that she didn't regret any of her one-night stands, describing herself as a true feminist and a strong woman who knows what she wants. She withheld her name, but she still didn't want the number of men she had slept with printed.

Same game? Another young woman found as a first-year student that nobody seemed to have boyfriends but felt, as she put it, "I can't just lose my V-card to some random guy." In the spring she picked a boy she had been dancing with and awarded her card to him. "I'm like, 'O.K., I could do this now. He's superhot, I like him, he's nice.'" Her expectations were very low. But because he let her spend the night and walked her home in the morning, all her friends were "super envious"; she came back with a huge smile on her face because she'd had "such a great first experience." She stayed on good terms with the boy and during spring break had casual sex with someone else. But she believes boys control the hookup culture: "It's kind of like a spiral." She explained that the girls stop anticipating that they will get a boyfriend, "but at the same time, they want to, like, have contact with guys. . . . [They] try not to get attached."

Same game? Haley, a senior, reminisced about going to frat parties as a freshman: she'd go in and they'd take her down to a dark basement. "There's girls dancing in the middle, and there's guys lurking on the sides and then coming and basically pressing their genitals up against you and trying to dance." These are dance-floor makeouts, or difmos. After one, she knew she was drunk and asked the boy to take her home; he took her to his room instead and had sex with her semiconscious body. She recalled another boy popping his head into the room: "Yo, did you score?" Eventually she looked back on this episode as rape.

Another girl said she usually ended up giving the boy oral sex, because by the time she got to his room she was sobering up and saw that as her best way out. Another, Kristy, told of making out with a boy in his house when he said, "'Get down on your knees.' . . . I was really taken aback, because I was like, no one has ever said that to me before." The boy said it was fair and, when she balked, he pushed her down. After that, she thought, "'I'll just do it.' . . . I was like, 'It will be over soon enough.'" Catherine, a senior, looked back on her hookups as a continual source of heartbreak.

I know those boys. I teach them. I have three daughters who had to navigate these rocky shoals in three different colleges, and they say they were able to keep the hookup culture at a distance. I can't really be sure, of course, but I do know that many do not. The young women in Taylor's story are not even remotely playing the same game, and if they think they are, they are way out of their league.

The research on this is clear, and transnational. A 2008 study by Anne Campbell was called "The Morning After the Night Before." A British television station surveyed thousands of people through its website; 998 of the men and 745 of the women who responded were heterosexual and had had a one-night stand. They were asked about their agreement with positive and negative statements about the event. Men were much more likely to report greater sexual satisfaction, well-being, and self-confidence, while women were much more likely to feel that they had been used and had let themselves down. Overall, subtracting negative scores from positive ones, men had more than double the net gain from the experience. As for regret, 23 percent of men but 58 percent of women said they would not repeat it.

Campbell writes of women's post-tryst distress, "The men had subsequently behaved disrespectfully and dismissively. . . . While not wanting a longer relationship, many women felt a strong sense of rejection"; they said they'd been "blanked." Women's positive remarks had to do with being made to feel sexy and wanted, craving male attention, or satisfying curiosity. One said, "I have a very poor self image and the man I slept with was a conquest. . . . He was very popular with other women and very good-looking. I thought that if I slept with him it would put me on a par with my prettier and more worthy peers. Unfortunately it didn't work and my self esteem/confidence suffered."

As for the sex itself, men spoke of "euphoria," "excitement and lust," and "blowing off sexual steam." Some women had fun and felt free, but most said things like

- "Thought it would be one of life's experiences, but it was nothing like the sex found in movies."
- "The expectation was better than the reality, the sex was rubbish."
- "The sex is never particularly satisfying because it is difficult to let go with someone you don't even know."
- "Not as good as sex with a partner; they are more into your needs and know your body a lot better."

Concerns like *Is sex all the person was after?* and *Will the person call or will they dump me?* were expressed by 81 percent of women and 17 percent of men. I can almost hear the men asking, *Dump you from what?*

Is there something about Britain or Campbell's methods that produced these results? Many U.S. studies confirm them with other approaches. Anthropologist John Townsend and his colleagues have studied hookups at Syracuse University for over a decade. In 2011 they reported on 335 male and 365 female students. Men were much more likely to endorse casual sex and women to feel a need for attachment and emotional involvement, but men were only somewhat more likely to have *had* such hookups. This suggests that men are persuading women to have more casual sex than they want while women prevent men from having as much as *they* want.

In 2012, Justin Garcia of the Kinsey Institute and his colleagues reviewed many studies of hookup culture, including patterns like NSA (no strings attached) sex and FWBs ("friends with benefits," a.k.a. "hookup buddies" or "fuck buddies"), which provide repeat encounters ("booty calls") with one person. In one study of 832 college students, 50 percent of men and 26 percent of women had positive emotional reactions after hookups, while for negative reactions the percentages were almost exactly reversed. In another, an online survey of more than twelve thousand students from seventeen colleges, 55 percent of first hookups involved oral sex only for the man and 19 percent only for the woman; 31 percent of men and 10 percent of women had orgasms. In a third study of 761 women, more than half reported at least one experience of unwanted sex. About two-thirds of hookups with vaginal intercourse involved condoms; when only oral sex was involved, the percentage was close to zero.

In "Bare Market: Campus Sex Ratios, Romantic Relationships, and Sexual Behavior," sociologists Jeremy Uecker and Mark Regnerus studied a representative sample of nearly a thousand single, heterosexual women in four-year coed colleges. The percentage of women on campus predicted attitudes and behavior toward the men on the same campus. Where women were plentiful, they were more likely to say that men were untrustworthy and uninterested in commitment. They expected less from men and found it harder to meet the right kinds of men. They were less likely to have gone on traditional dates, to have had a boyfriend, or to be a virgin and more likely to have had sex in the last month, especially if they *didn't* have a boyfriend.

Little wonder that women on those campuses say they don't want their percentage to go up any further. By the way, if women wanted casual sex as much as men, wouldn't they be *more* likely to have it where they are in a minority? They don't, but where they predominate the market leads them to give up more than they otherwise would. They are competing for men, and the men have more power. Since women now make up about 60 percent of college students and rising, it seems likely that hookups will rise as well. The National Longitudinal Study of Adolescent Health has found a similar effect of sex ratio in high schools, which suggests that the next cohort of college women will be playing the hookup game as they continue to try to please scarce men.

I recognize that many women speak well of these experiences and often seek them. Women like sex, and some like casual sex. Among other things, they are protecting themselves from relationships that might distract them from their studies and career pursuits. Celibacy is often not a good choice, especially given peer pressure and boy pressure, not to mention desire. If you want to avoid being seen as a child or a prude by women and you like men enough to want to be on good terms with them, hookups can be stimulating and convenient. One time in ten or so, you may even have an orgasm. A so-called fuck buddy can be a reasonable compromise between loneliness and a complicated romantic involvement. But these encounters and relationships are sexually asymmetrical. Someone is often being used, and it is seldom the boy.

Recall that Nisa, the !Kung woman who was the subject of my late wife Marjorie Shostak's classic, said, "Women possess something very important, something that enables men to live: their genitals." Or as anthropologist Donald Symons, whose pioneering book *The Evolution of Human Sexuality* helped start this field of research, said, "Among all peoples it is primarily men who court, woo, proposition, seduce, employ love charms and love magic, give gifts in exchange for sex, and use the services of prostitutes. And only men rape. Everywhere sex is understood to be something females have that males want." This is an exaggeration but one with a great measure of truth. Women forgot this truth when men convinced them that both sexes have the same interests; this was the sexual revolution of the sixties, which involved a lot more change and disappointment for women than for men, at least in the realm of sex. In view of persistent myths about this in the ongoing sexual revolution today, let's look at the evidence for Symons's pointed claim.

A 2001 overview in the *Personality and Social Psychology Review* by Roy Baumeister, Kathleen Catanese, and Kathleen Vohs combed more than 150 studies to answer the question "Is there a gender difference in sex drive?" Overall in these studies,

> Men have been shown to have more frequent and more intense sexual desires than women, as reflected in spontaneous thoughts about sex, frequency and variety of sexual fantasies, desired frequency of intercourse, desired number of partners, masturbation, liking for various sexual practices, willingness to forego sex, initiating versus refusing sex, making sacrifices for sex, and other measures.

There were *no* studies with contrary findings—not a single one indicating stronger sexual motivation in women than men. In one typical study, 90 percent of men but only half of women felt sexual desire at least a few times a week. In another, the average young man was sexually aroused several times a day, the typical young woman "a couple of times a week." In an Australian survey, people who were in a committed relationship, *wanted* to have sex, but were *not* having it were almost exclusively male.

Compared to women, men begin to have sexual intercourse earlier in life (despite later puberty), are less willing to give up sex for any part of life, are more permissive and favorable toward sex, initiate sex much more often in longer relationships, and show more preference for every sexual practice, including, rather astoundingly, cunnilingus. Although men often have physical problems like premature ejaculation and erectile dysfunction, hypoactive sexual *desire*, whether by diagnosis or self-report, is overwhelmingly female.

In their classic 1989 study, "Gender Differences in Receptivity to Sexual Offers," social psychologists Russell Clark and Elaine Hatfield had confederates who were college men and women of average attractiveness approach strange but attractive members of the opposite sex on campus and say, "I've been noticing you around campus lately and find you very attractive. Would you like to go to bed with me tonight?" Of the men, 75 percent said yes; of the women, 100 percent said no. Some men said things like "Why wait till tonight?" Some women said things that I won't repeat here.

In 2011, psychologist Terri Conley claimed to have repeated this study; she found that under some conditions, the difference between the sexes was much smaller. However, there was a minor problem with this "replication": it was purely a paper-and-pencil study. I am sorry, but that is not remotely a repeat of the classic study. Yet the media seized on it as proof that sexual mores have dramatically changed and that women are almost as interested in casual sex as men. The current wish to deny the facts of life is very great.

Suppose we ask what happens when you remove the slight complication of having to deal with another person, whether a stranger, intimate long-term partner, or anything in between. Sex differences in masturbation are consistent and large. Women are much more likely to have never masturbated; women who do masturbate do it much less frequently at all ages than men. In a 2011 summary of meta-analyses and large data sets, psychologists Jennifer Petersen and Janet Shibley Hyde confirmed substantial sex differences in masturbation—even with such blunt (and indeed almost ridiculous, if you are looking at sex differences) measures as whether someone has masturbated in the past *year*—and in pornography use, in the usual direction.

Consider people's fantasy lives. Here, too, there is no other person present. You can dream up whatever you want, no risk, no compromise, no complications. In one study, Bruce Ellis and Donald Symons gave three hundred male and female students an anonymous questionnaire. Men (32 percent) were four times as likely as women to say they had fantasized about having sex with more than one thousand different people (by *college* age). Men were much more likely to say that visual images were more important than touching in their fantasies (66 percent versus 39 percent), women twice as likely to say touching (55 percent versus 28 percent). Men were about twice as likely to focus on visual images rather than feelings, women three times as likely to say feelings. And men were almost three times as likely (48 percent versus 17 percent) to agree that in their fantasies, "the situation quickly includes explicitly sexual activity." These findings have been repeated in many studies.

Of course, sometimes men want men, and women, women, and these relationships are most instructive. Lesbian relationships, compared to those between gay males, are less sexual at every stage by almost every measure; the phrase "lesbian bed death" may be an exaggeration, but we don't often hear it said about gay men. In many ways—frequency of intercourse, open relationships, sexually transmitted diseases, and the use of sadomasochistic elements in sex—heterosexual couples are intermediate between lesbian and gay male pairings.

These are old findings, and new research holds few surprises. A 2013 study by sociologist Bethany Everett, as well as research in 2010 by epidemiologist Fujie Xu and colleagues, showed that while bisexual people are at the highest risk, the number of lifetime partners and the amount of sexually transmitted disease are higher in exclusively gay than in heterosexual men and lower in exclusively lesbian than in heterosexual women. So when you remove such complex issues as male dominance and women's oppression or the desire to please in heterosexual relationships, *where the two people involved are of the same sex*, male-female differences are larger, not smaller.

Who pays cash for sex? Almost exclusively men. An estimated one-tenth to one-sixth of U.S. men have paid for sex, half of those while they were involved in other relationships. A few women pay for sex, but it's not just a simple transaction. Women do not buy much pornography or leer at pictures of naked men. Porn customers are overwhelmingly male, and the main counterpart for women is romance fiction. In her insightful 2012 article "The Pop Culture of Sex: An Evolutionary Window on the Worlds of Pornography and Romance," psychologist Catherine Salmon says, "Romance and pornography are both multibillion dollar industries, and their stark contrasts reflect the deep divide at the heart of male and female erotic fantasies."

After an extended analysis, she concludes, "Pornography is a male fantasy world of short-term mating success while the romance is a female fantasy world of long-term mating success. At their hearts, that is what they are, fantasies that are reflections of the different ancestral problems faced by males and females in the mating domain." Even the huge hit novel *Fifty Shades of Grey,* sympathetically reviewed by some intelligent women and read by millions, is a romance novel that includes soft-core pornography. It depicts a rich and powerful man deeply in love with a younger woman and very concerned about her welfare and her sexual pleasure. The bondage and discipline he subjects her to (because of his self-described mental illness and after elaborately and solicitously gaining her consent) is mild compared to that in male-directed pornography; she is depicted by the (female) author as not hurt but enraptured, and the one time she is really hurt, she leaves him.

Symons's remark in his book about what women have and men want implied that heterosexual sex is a scarce resource for which men strive, compete, and pay; he repeated it to a TV interviewer in a singles' bar, where men are more likely to pay for the drinks and where there is no such thing as a "men's night"—letting men in for free to attract women as paying customers. Roy Baumeister and Kathleen Vohs summarized the broader evidence about what is clearly a market in another paper, "Sexual Economics." Most prostitutes are women who overwhelmingly serve male customers, and almost all male prostitutes serve men as well. Women rarely pay for sex, and when they do it often involves the pretense of a romantic tryst; they demand something more than a simple hookup—which is what men are typically paying for. When the famous actor Charlie Sheen was found to have paid top dollar for sex with one of the so-called Hollywood Madam's young women, many wondered, *Why would a man who could have his pick of willing women for free pay $1,500 (about $2,500 in today's dollars) to have one come to his room?* Answer: *He doesn't pay her to come to his room. He pays her to leave.*

Strip clubs overwhelmingly consist of women performing for mainly male customers, although some women attend, often with male dates. There are about four thousand of these clubs in the United States, employing some 400,000 women. The reverse situation, in which men strip for women, represents a very small fraction of these numbers. There are also strip clubs for gay men, but very few for lesbians.

Finally, coercive sex is overwhelmingly male; 99 percent of FBI arrests for rape are of men. It is not obvious that it has to be this way because males have to be aroused to have sex; women could force men or other women to give them oral sex under threat of violence, as men do with both male and female victims, or use dildos or other objects to rape men anally, to humiliate them, as men do to victims of both sexes. Such assaults by women are vanishingly rare. A small percentage of gang rapes of women involve both sexes, and a small percentage is perpetrated by groups of women. Rape occurs in some lesbian relationships, but with nothing resembling its frequency in heterosexual and gay male relationships.

These are sound generalizations, not absolute rules. Some women do want sex as much as any man. Some men want little or none. There is nothing inferior about wanting it and nothing superior about not wanting it—although (despite the substantial minority with hypoactive sexual desire) women certainly have the potential for superior orgasmic capacity. But denial of the facts of human sexual nature as it applies to most men and women can only lead to confusion and, ultimately, to suffering. Male sexuality is *driven*. Men frequently want sex, period, while women tend to prefer it in the context of a relationship, a physical connection allied to an emotional one. Regardless of what I may privately desire, and regardless of how natural men's needs may be, I can't see that those divergent preferences are equally admirable.

To think that all these differences could result merely from cultural arrangements is naïve in the extreme. We now have overwhelming evidence that Elizabeth Cady Stanton's "difference in man and woman"—by which she meant behavioral, psychological, and moral dispositions—is in part grounded in biology. Scientists once reticent in their assertions have become very bold. In 2011 the journal *Frontiers in Endocrinology* brought leading experts together. Psychologist Melissa Hines, for decades a respected researcher on the effects of prenatal hormones on gender, reaffirms their power but allows for two other newly proven influences: direct genetic effects on the brain, and the intrauterine environment. Neurobiologist Margaret McCarthy agrees that genes as well as hormones matter. She adds new evidence for how male and female become different in the hippocampus and amygdala—parts of the emotional brain outside the hypothalamus. Gender psychologists Sheri Berenbaum and Adriene Beltz show how exposure to high levels of prenatal androgens masculinizes later activity and occupational interests, sexual orientation, and some aspects of spatial ability, and they also find that pubertal hormones appear to influence gender identity and perhaps some male-female differences in psychiatric illness.

Neurobiologist Ai-Min Bao and her colleague Dick Swaab reviewed growing evidence for sex differences in the *human* hypothalamus and found that male-to-female transsexuals resemble women in these measures. They believe that male-female differences in cognition, gender identity, sexual orientation, and neuropsychiatric disorders are "programmed into our brain" very early on and, remarkably, that "there is no evidence that one's postnatal social environment plays a crucial role in gender identity or sexual orientation." Simon LeVay, who wrote the introduction for the special issue, had an established reputation as a neuroscientist studying the visual system when he came out to the world after showing that in one hypothalamic area gay men resemble women and differ from heterosexual men. Two decades later, he reconsiders the idea that social experience influences sexual orientation and gender identity, concluding that "little direct evidence supports this notion at present."

The work is ongoing. Julia Sacher and her colleagues, in a 2013 summary of brain-imaging studies, found differences beyond the hypothalamus. Controlling for brain size, women have more gray matter and a thicker cerebral cortex. Women show stronger connectivity in the left brain, men in the right, contradicting expectations about men's logic and women's emotionality. Some studies find differences in the corpus callosum, a huge highway across the brain, which could mean the hemispheres collaborate better in women. Many studies show brain differences in activation of particular circuits in emotional and mental states, although this need not be causal. In 2014 Amber Ruigrok and her colleagues reanalyzed 126 brain-imaging studies and found several anatomical differences, although even such structural differences could theoretically result from upbringing. However, Sacher also showed in 2013 that there are many functional changes in brain activity over the menstrual cycle.

It's important to understand that the *similarities* between men and women's brains are much greater than any differences; the differences that exist are unrelated to general intelligence, but they are tied to specific dispositions. A key finding is that the male amygdala is relatively larger and dotted with testosterone receptors, while the prefrontal cortex, which inhibits aggressive and other impulses coming from the amygdala, is larger and develops earlier in women. These differences, combined with hormonal effects on the prenatal hypothalamus, could help explain why men greatly exceed women in violence and driven sexuality.

Incidentally, it used to be said that women could not be airline pilots or heads of state because of the supposed emotional swings of the menstrual cycle. That was before a landmark study, "Body Time and Social Time," by sociologist Alice Rossi and economist Peter Rossi, the first and perhaps the only major menstrual cycle study that included men. For one thing, weekends had a much bigger impact than cycle phases. More importantly, men had the same number of bad days a month as women, except that the women's were cyclical. So, would you rather have your airliner or your country piloted by someone who has bad days at random or someone who has the same number of bad days coming around like clockwork?

As for women being more emotional in general, we have seen that it depends on which emotions we are talking about. Women cry more easily, but male politicians tear up quite frequently in public. Women show more empathy in most situations, but men are far more likely to have violence on a hair trigger in international relations, and to feel and succumb to inappropriate and even destructive impulses while in office. Men are far more often distracted by sexual impulses and fantasies, even if these don't result in problematic behavior. And if egotism and exaggerated ambition are emotional, which sex has more of those? All in all, the world will become safer and more efficient as women take their proper roles in leadership.

*Chapter 9*

✦

# Developing Daughters

Men have ceded some power to women partly because we have daughters and want them to succeed. It's not that we don't love our boys, but girls have been kept down for thousands of years. I have a son, two daughters, and a stepdaughter, and I want them all to do well. Boys today are running into serious trouble, and they need certain kinds of protection. But in terms of who runs things, in the present as well as the long course of history, the odds remain stacked against girls. So this chapter's title has two different meanings: it's both about how girls develop into women and about how *we* can develop *them* into great achievers without making them mimic boys and men.

Journalist Hanna Rosin's provokingly titled book *The End of Men* made the case that among younger women, a replacement process is well under way. I agree—but anthropologists take the long view. When Elizabeth Cady Stanton made her stunning speech about the difference, a devastating war had just freed the slaves, but the nonviolent struggle for women's rights was only beginning. Half a century later women would vote, control property, find out about and use barrier methods of birth control, and begin to infiltrate the professions.

Half a century after that, women realized how far they had to go and created second-wave feminism. At stake were equal pay for equal work, equal access to education, the media, the military, and the professions, new and full reproductive rights, sexual freedom of choice, the end of the double standard of sexual morality, a real voice in the public square, the option to escape from an abusive or otherwise bad marriage, an end to rape and other violence that kept women down and out, and a final expiration of *Kinder, Küche, Kirche.*

Now we are closing in on yet half a century more, and we should take stock. Rosin's hard-hitting book was criticized by many for predicting male demise, of course, but not all women's advocates welcomed it, either. Rosin, they said, grossly exaggerated progress. I don't think she did. Although I don't want to see the end of men, I think progress for women is now steady and irreversible. Rosin reasonably looked back as well as forward, but on a time scale of decades. In this book I have asked you to look back millions of years, then thousands, then hundreds; we can look forward generations at least, but even centuries and millennia start with decades.

I acknowledge the limitations. As of 2014, women were CEOs of only 4.8 percent of Fortune 500 companies and 5.2 percent of Fortune 1000 firms. They held only 78, or 18 percent, of the 435 seats in the U.S. House of Representatives and only 20 of the 100 Senate seats. Women headed only 20 of some 190 nations. Only 26.4 percent of colleges and universities had women presidents. This looks bad. But with one exception—there were also 20 women heads of state when Rosin wrote—every one of these numbers was better. *In two years.* Not one of them slid back.

The CEO percentage is the smallest but the fastest rising. It was 3 percent in 2011. That's a 60 percent increase in less than the time it took for one of my kids to go through middle school. And by the way, women are not just running cosmetics, apparel, and food corporations. They've been running General Motors, Hewlett-Packard, IBM, Archer Daniels Midland, Lockheed Martin, DuPont, General

Dynamics, Oil States International, Xerox, Duke Energy, Gannett Company, Yahoo, Alliant Energy, Schnitzer Steel, ITT, International Game Technology, Clearwater Paper, and Benchmark Electronics. Boy stuff—the machinery of the world. And on average, *they outearn males* on the same list. On May 9, 2014, Susan Story became CEO of American Water Works Company, the largest water and wastewater utility company in the United States. What is more, the proportion of women in the next tier, ready for *future* CEO positions, is 14 percent, and there are still more in the next tier after that. CEOs are not usually young, and most started in a time when barriers to entry for women were much higher. This means, in my view, that the future is wide open.

By the way, since corporate boards are overwhelmingly male, men must have voted to make women the CEOs of all those companies, just as male legislators had to vote to give women the same right. Politics is complex, but in these massive companies the boards were not appointing women out of the goodness of their hearts; nor were they risking billions of dollars for political correctness. They were picking the best person.

Two of the companies mentioned above, Lockheed Martin and General Dynamics, are mostly devoted to defense. Does this negate the idea that increasing numbers of women leaders will make for a more peaceful world? Not at all. Women have every right—in fact, they have a responsibility—to play a role in defending their country, including joining the armed forces, fighting, rising in rank in the services, and becoming leaders there and in the military-industrial complex. The positive influence of women on the safety of the world is a matter not of their avoiding certain professions but of the added stability they and their priorities will bring to whole societies benefiting from their judgment and their leadership.

But how should we think about the percentages? How are women faring across the board in becoming leaders? Is the Senate glass 80 percent empty or 20 percent full? How about the CEO glass—95 percent empty or 5 percent full? And the head of state glass—90 percent empty? Having lived from the 1940s until now, I have absolutely no doubt about the answer. What would my mother say, who never saw a woman delivering the TV news, being asked for an expert opinion, or sitting in a cabinet meeting? Or my European immigrant grandmothers, who couldn't vote until they were about forty years old and for whom the idea of a woman prime minister of Britain or Germany would have sounded like a hurtful joke? What would *they* say to my daughters—their great-granddaughters—about the changing prospects for women?

When I went to college, in the 1960s, professional schools had a handful of women in an entering class of one hundred. Today women make up 47 percent of medical school classes, and law schools are similar. More than 40 percent of students entering MBA programs—the pool of future CEOs—are women. Women are 30 percent of federal district court judges, 32 percent of federal appeals court judges, and one-third of the U.S. Supreme Court.

Of course, there is a long way to go, and women *should* be impatient. If not for their impatience, I would be much less optimistic. But change has been happening steadily for 150 years in the Western world. The questions are: How do we make it happen faster here? And how do we help the rest of the world catch up?

I happen to live in Georgia, not a bastion of women's rights. A young friend of mine, Anna Beck, just served four years as executive director of Georgia's WIN List—"WIN" stands for Women in Numbers. The organization puts pro-choice women in office in the state by campaigning and fund-raising but also by training a cadre of young women as future candidates who will know what it takes to win. During Anna's four years alone, Georgia's WIN List protected 96 percent of women incumbents in 2010, added four new women to the statehouse in that year, protected 90 percent of women incumbents in 2012, and added four new women to the statehouse in 2011 and 2012. WIN's programs have been mounted in a hostile male environment in one of America's least educated and least feminist states, in the face of redistricting that favors right-wing male candidates, and amid initiatives to roll back reproductive rights. Grass-roots organizations like this are emerging in many places. They are holding the line for women's rights, and they are advancing women's power.

Nationally there is EMILY's List, named for a saying: *Early Money Is Like Yeast*. By raising more than $50 million for 2012 U.S. Senate campaigns, this PAC helped incumbent Claire McCaskill of Missouri beat a man who had claimed that rape victims can physiologically prevent pregnancy. It elected Elizabeth Warren, a Harvard Law professor and consumer leader, who defeated the popular, handsome, gun-loving male incumbent by a comfortable margin; Mazie Hirono of Hawaii, the first woman senator from her state as well as the first Asian-American and the first Buddhist in the Senate; and Tammy Baldwin of Wisconsin, the first openly gay person ever to sit in that body. These victories created the largest class of Senate women in history.

In 2012 EMILY's List also helped elect Maggie Hassan, the only pro-choice Democratic woman governor in the country, as well as nineteen new women to the House and five Senate incumbents besides McCaskill. EMILY's List was founded in 1985 with twenty-five members; they soon helped elect Barbara Mikulski, who brought the number of women Senators to two. By now, they have helped bring one hundred women to the House, nineteen to the Senate, ten to governorships, and more than five hundred to state and local office.

President Barack Obama, on the fiftieth anniversary of Martin Luther King Jr.'s "I Have a Dream" speech, paraphrased King, who'd said, "The arc of the moral universe is long, but it bends towards justice." Obama said it "may bend towards justice, but, it doesn't bend on its own." Hillary Clinton, in conceding the 2008 Democratic presidential nomination to Obama and endorsing him, said that as a woman, she was well aware of the biases still in place. But she also said she ran as a daughter whose mother never dreamed of such opportunities, and as a mother worried about her own daughter's future. It so happened that at that moment, the fiftieth woman to leave the planet for outer space was orbiting overhead. "If we can blast fifty women into space," she said, "we will someday launch a woman into the White House. Although we weren't able to shatter that highest, hardest glass ceiling this time, thanks to you, it's got about eighteen million cracks in it, and the light is shining through like never before." But she also looked back to the suffragists, abolitionists, and civil rights heroes. Because of them, she said, she grew up taking it for granted that women could vote, and her daughter took it for granted that schools could be multiracial. "Because of them and because of you," she continued, "children today will grow up taking for granted that an African-American or a woman can, yes, become the president of the United States."

In the summer of 2013, three years before the 2016 presidential election, EMILY's List held its first "Madam President" town hall in the early primary state of Iowa. The event's organizers say they intend to hear that phrase in the White House in 2017 whether or not Hillary Clinton runs. They are already investing in that outcome. Early money is like yeast, and this bread is rising.

But a book about the whole sweep of evolutionary history must look at the world, not one country, and at a broader range of political views. Angela Merkel (a former quantum chemist) is chancellor of Germany, the population center and economic powerhouse of Europe; recently elected to a third term, she will be one of the longest-serving modern European heads of state. Norway and South Korea, both also leading democracies and economic powers, are headed by women, as are Brazil and Argentina, the two giants of South America. (By September 2014 Brazil's leader was facing one serious challenger—also a woman.) India, the world's largest democracy, was led by Indira Gandhi from 1966 until (except for 1977–80) her assassination in 1984. The prime minister of Britain from 1979 to 1990 was Margaret Thatcher, who spearheaded dramatic change as she presided over a country pivotal in the Atlantic alliance and the Commonwealth of Nations. Julia Gillard, as prime minister of Australia, protected a key Pacific economy from the Great Recession; Australia fared better than almost any other advanced nation. And Israel was led by a woman prime minister, Golda Meir, during some of the most difficult years of her country's history.

Of course, there are many skeptical things to be said about these women: how few they were, how they made it in a man's world by mimicking men, how they failed to protect women's interests, how they succumbed to war. But what cannot be said about them is that they were incompetent or that they proved a single one of the claims made for centuries—involving emotions, cycles, weak constitutions, delicacy, whatever—about why women cannot lead. *They proved the opposite.* They did it. Warts and all, they performed as well or better than men in the same positions before and since, and so will many more women after them.

Incidentally, the new Norwegian prime minister is the second woman in that position in the country's history, and the two came from opposite sides of the political divide. Two of the three top cabinet positions went to women, and in recent years Norwegian governments have hovered around half women. It's interesting to recall that these women, and the people who elected them, are descended from the Vikings, one of the most brutal male-dominated cultures in history; Viking men used rape and other violence against conquered women from Ireland to Russia. Now women run their country.

But as much as all this reveals about women's potential, it is not the human species; it is merely the top of the top.

In June 2013, a three-year-old organization called UN Women published a declaration: *A Transformative Stand-Alone Goal on Achieving Gender Equality, Women's Rights and Women's Empowerment.* UN Women merged four existing UN divisions that dealt with women's issues. The "transformative stand-alone goal" is key; there are overall Millennium Development Goals (MDGs) for all humanity. But now there is a stand-alone goal for women, because while the MDGs include women's concerns, more focus is needed. The report concludes,

> These ideas are not new. . . . Yet, addressing them in a holistic and comprehensive manner . . . would constitute a ground-shift in development policy and practice. . . . The world simply cannot afford to miss this once in a generation opportunity to transform the lives of women and girls and men and boys everywhere.

There are three components to the declaration: (1) freedom from violence against women and girls; (2) gender equality in capabilities and resources; and (3) gender equality in decision-making power. Each component has between three and eight targets, and each target is monitored according to several quantitative measures; every member country (and its associated UN personnel) will be expected to submit these numbers and show progress on them every year.

These are ambitious goals. Many countries cannot even accurately count births and deaths, much less measure so many subtle indicators. However, UN Women is monitoring its employees throughout the world to ensure quality control in these reports. It is good to have specific targets and start trying to measure progress toward them. And it is hugely important that women are being singled out, and not just for their sake.

We have long known that the best way to spend a development aid dollar is on educating a girl and, thus, empowering a woman, because it is proven that she will:

- have fewer births and lower infant mortality;
- raise healthier children of both sexes;
- educate those children to higher levels;
- have better health herself;
- improve the health of her husband;
- have more lifetime earning power; and

- better resist male attempts, through threats, abuse, and violence, to divert her time and money from these aims.

So, by focusing on half the human species, UN Women is better serving the broad development goals of the other half as well. It will help guide the next phase of human evolution.

Meanwhile, the indicators show what is at stake. While the third component, gender equality in decision making, may not differ so much between the developing and the developed world, most readers of this book are not worrying about sanitation, finding toilets, access to obstetric facilities, pressure to have their daughters' clitorises cut out, or girls risking rape, disfigurement, or death for going to school. Domestic violence is all too common in all countries, but in some, routine violence by male partners is the central fact of life for women. We should understand these scourges but, more importantly, what is working against them. There is no reason to despair, but there is a need to actively bend the arc of history; do that and the arc of evolution will take care of itself.

During evolution, we adapted to a greater male mortality by supplying (biologically) a greater proportion of males at conception; this continues to be true at birth. In the European Union, for example, that proportion is around 1.06, meaning 6 percent more newborn boys than girls. Taking all children under fifteen, it is slightly lower, 1.05. In adults it is equal (about 1.0), but in the years over age sixty-five it is reversed, with only 75 men for every 100 women (.75). These numbers reflect the basic biology of our species under conditions of good nutrition, sanitation, health care, and relative gender equality.

In China, however, the corresponding numbers are 1.12, 1.17, 1.06, and 0.93; in India, they are 1.12, 1.13, 1.07, and 0.9. Some other countries are similar. Something is wrong here—at every age there are more boys and men than expected from basic biology—and together these countries contain a third to a half of our species, so it is not just strange but important, and we know what is at work. Girls and women are discriminated against beginning in the womb, where they are much more likely to be deliberately aborted if their sex is known. But the ratio worsens in childhood, meaning more girls than boys die, because girls are neglected. After age sixty-five, there are about 20 more men per 100 women in India and China than in Europe. This means that women, despite being naturally resilient, are dying in greater numbers throughout life. In China, the one-child policy worsens this gender bias.

Amartya Sen, a Nobel laureate economist, has called them the missing women; there should be 100 million more than there are right now in the developing world, 85 million just in China and India. The result: many millions of men without women, which should favor women's power but usually leads instead to male violence, rape included. Evolutionary biologists know that in most species, skewed sex ratios eventually right themselves, because the scarcer sex does better in the mate market, but waiting for this rebalancing takes time and pain.

Kate Gilles and Charlotte Feldman-Jacobs of the Population Reference Bureau, in a 2012 policy brief, recommended discouraging the use of sex determination technologies (SDTs) for the unborn, recruiting doctors as partners and advocates, giving women rights of inheritance, and mounting awareness campaigns. Various countries, including China and India, have restricted SDTs; India has a "Doctors for Daughters" campaign. The Ladli program in Delhi puts money in a bank account on the registration of a girl's birth, with further deposits rewarding school progress; girls get the money at eighteen if they have stayed single and finished tenth grade. A TV series called *Atmajaa* (Born from the Soul) dramatized the plight and value of girls and women in northern India, where the imbalance is worst; research showed that the show changed younger women's attitudes.

The biggest success story is South Korea, which has left the ranks of skewed-sex-ratio countries, after being worst. South Korean officials enforced severe penalties for physicians offering sex selection, and they campaigned to change patriarchal attitudes and give women well-paying jobs. Between 1985 and 2003, women who said they "must have a son" declined from 48 to 17 percent. Today, if parents express a preference, it is more likely to be for a daughter. World Bank demographer Monica Das Gupta, a bit stunned at the speed of the change, has said that son preference in South Korea "is over."

Another unexpected change is the plummeting birth rate in Bangladesh. Ruth Levine, Molly Kinder, and the "What Works?" Working Group made it a case study in *Millions Saved: Proven Successes in Global Health.* The program relied on a large cadre of women outreach workers, a small army of Margaret Sangers giving out advice, education, and birth control, supported by mass media. Contraceptive use went from 8 to 60 percent in the last quarter of the twentieth century, while births per woman dropped from more than six to three; at last count, in 2012, the number was 2.2.

A 2012 multivariate analysis showed that education and the media made most of the difference, although religion also mattered—Hindus use birth control more than Muslims. Top writers created a soap opera centered on Laila, a family-planning outreach worker; the show changed husbands' attitudes and reduced harassment of real outreach workers. The percentage of married women using birth control in 2007 increased with age, from 41 percent for those in their late teens to 67 percent in their late thirties; mature women had learned to stop bearing children and focus on those they have. This initiative is one of the great triumphs in family planning, and it is being imitated in Kenya, Tanzania, Brazil, Mexico, and India.

But the births that occur remain dangerous. Worldwide, 287,000 women die of childbirth-related causes every year, 99 percent of them in developing countries. A woman in Afghanistan or Sierra Leone has a lifetime risk of dying in childbirth that is 1,000 times higher than that in Norway or Switzerland. Yet Sri Lanka has had dramatic success, halving maternal deaths every twelve years since 1935 and going from around 600 to 60 deaths per 100,000 births since 1950—this with a per capita health-care budget half the size of India's, where *400* mothers die per 100,000 births. Sri Lanka is poor and was torn for a quarter century by an agonizing civil war that didn't end until 2009. How could it defeat death in childbirth?

First, Sri Lanka has a tradition of female literacy; 89 percent of women can read and write, more than twice the South Asian average. It has had two women prime ministers, although only 5 percent of Parliament is female. It has good civil records, logging all maternal deaths since 1900; widespread and growing access to free health care; training of midwives (more than 97 percent of births are now professionally attended); outreach to poor, isolated women; and an all-out war on malaria. The Sri Lankan case, along with less dramatic declines in Honduras, Malaysia, and elsewhere, shows that the key variables—aside from educating girls—are government initiative (including good statistics), hospitals, and well-trained birth attendants in the community. Incidentally, 99 percent of Sri Lankans say religion is important in their lives; 70 percent are Buddhist, 13 percent Hindu, 10 percent Muslim, and 7 percent Christian.

My colleague Lynn Sibley—an anthropologist, nurse-midwife, and professor of nursing—illustrates what one determined person can do. "The things that kill mothers and babies happen quickly, they're emergencies, they can't be predicted," she says. Having delivered thousands of babies, she should know. But in Ethiopia, where she has been working for years, 90 percent of women today give birth without that expertise. Traditional birth attendants (not trained midwives) assist them in their homes in rural villages, and maternal deaths are frequent.

Sibley identifies a forty-eight-hour period during and after birth when women are most vulnerable—to hemorrhage, eclampsia (a blood-pressure crisis), obstructed labor, and sepsis (systemic infection); these things kill mothers in Ethiopia and throughout the world. But to her those forty-eight hours are a window of opportunity. Ethiopia, a country of 80 million, has the world's ninth-highest birth rate, with 720 maternal deaths per 100,000 births, or twelve times the rate in Sri

Lanka. When the Bill & Melinda Gates Foundation, which gets the best advice there is about global health, heard about Sibley's program, it solicited a grant application and soon gave her $8 million.

She knew what to spend it on. Take hemorrhage, for example. When a woman who has just given birth is (as we used to say in medical school) "trying to bleed to death," there are ways to "talk her out of it"—IV oxytocin; blood transfusions; in the worst case, hysterectomy. But if she "tries" this in a mud hut in the Ethiopian mountains, with a traditional village birth attendant, she may well succeed in dying.

Sibley's idea was to teach local birth attendants a few simple things: First, recognize and respond to excess bleeding; her research showed that they did not do this before training—some even had superstitions about it—but knew the signs very well after training. Another was getting the mother to urinate; a full bladder can inhibit the contractions that stop the bleeding. (Birth, I also learned in medical school, is not a clean process, but mothers are inevitably embarrassed by some of the most common things.) Another was massaging the uterus from outside to make it contract. Yet another (after the delivery) was combined pressure from a hand or fist inside the vagina and the other hand pressing down on the uterus from outside. Finally, they were taught to get the mother to a medical center right away.

These are simple, elegant interventions that (up to a point) respect local customs and grow out of alliances with local people; Sibley's program can, and I believe will, be scaled up throughout the poorest parts of the world. As she puts it, "Basically, no woman should die giving birth, knowing what we know today."

Thirty years ago, few people predicted that a strange disease limited to certain populations would become a great burden on the world's women and men. By the end of the 1980s, Thailand topped the charts in the HIV/AIDS epidemic but also soon led in fighting it. At first drug injectors and men who have sex with men prevailed in the epidemic, but HIV positivity in Thai female sex workers went from 3.5 to 22 percent between 1989 and 1991; among twenty-one-year-old male army conscripts it went from 0.5 to 3 percent in the same two years. Then the war on AIDS was brought into the prime minister's office, and a frank, intense campaign reduced new HIV infections by 90 percent in twelve years. Health workers visited brothels, where they entertained and taught; wooden pop-up models of penises (a red dot on the tip meant infection) made the young women laugh, even as they learned to slip condoms over the makeshift shaft. Public media campaigns directed at men did the rest.

According to the *Bangkok Post* of December 21, 2012, Mechai Viravaidya, who led the campaign and is known as "Mr. Condom," now warns, "The government has fallen asleep at the wheel." New infections are up again, threatening the general population. But in the 1990s, his "100 percent condom program" made all the difference (despite not getting that perfect score); the World Bank estimates that 7.7 million infections were prevented in Thailand by 2005.

Yet, as pointed out by Elizabeth Pisani, the brilliant epidemiologist and author of *The Wisdom of Whores,* something else was going on in Thailand in the 1990s. Young unmarried women were having much more sex—a tenfold increase over the decade. If you are scratching your head, I was too. How does more women having sex translate into less HIV? What has happened in country after country is that as young women became more sexually liberal, the percentage of young men paying for sex dropped precipitously. So, paradoxically, the last thing you want is a "Just say no" campaign for ordinary teens. And it's not a no-brainer to figure out how to subvert this epidemic or how to protect women in particular.

Today, for humanity as a whole, the proportions of men and women with HIV are even. But since women represent 60 percent of new cases worldwide, they will soon surpass men in total cases. Sex workers make up a small fraction; most women get it from their husbands. Men bring it to women, who get it more easily; the vagina is a cozy spot for germs. The heterosexual epidemic is worst in sub-Saharan Africa, but many men throughout the world decline to use condoms, and many coerce their wives to have sex on their terms. In Africa, public health campaigns in some countries, notably Uganda, were extremely successful for a time, but as in Thailand, there is backsliding.

According to a 2012 survey in Uganda, condom use is inadequate and new infections in that country are now rising. Meanwhile, in Madagascar, Mozambique, Zimbabwe, and elsewhere in southern Africa, some men want not only condom-free but "dry sex." This means putting abrasives or other substances in the vagina to increase friction; these widespread practices raise the likelihood of vaginal microtearing, which raises the risk of HIV and other sexually transmitted infections. As "Coca Colo" put it in her *Feminomics* blog,

> Dry sex aids are the inverse of lube. The idea is that it makes sex feel hot, tight, and rough, and presumably increases the enjoyment for the man. In practice, it can also make sex extremely painful for women, create vaginal swelling and peeling, and lead to a reduction in the body's natural protection from HIV.

Coca's points are consistent with research. However, men's infidelities and sexual pressures, although important in the African AIDS epidemic, did not start it. At the root, experts are finding, are health-care personnel. Today and since before the epidemic began, health workers in Africa, lacking resources, have reused unsterilized needles and surgical tools on patient after patient. This, along with the fact that the virus evolved in that continent, may be what initially made Africa the worst place in the world for HIV/AIDS.

So women need protection from incompetent health workers as well as pushy men. Bill Gates, as impressive at global health philanthropy as he was at software development (partly because of his wife Melinda's counsel), assessed their foundation's progress in 2012. He doubts the enduring success of behavioral change but writes,

> A second approach is male circumcision, which reduces HIV transmission by up to 70 percent. . . . The cost is quite low and the protection is life-long. Over 1 million men ages 15–49 have been circumcised in 14 Southern and Eastern African countries with large AIDS epidemics, but that is only 5 percent of the total number who could benefit from the procedure. . . . Botswana, Kenya, South Africa, and Tanzania are starting to show leadership by getting the message out to all young men. . . . Kenya has made the most progress, circumcising 70 percent of eligible men. I will be very disappointed if, by 2015, any fewer than 15 million young men have chosen to protect themselves and their partners by getting circumcised.

Notice the word "chosen." This is voluntary. Yet there's a certain poetic justice in a man showing up to have his penis clipped as a way to curtail an epidemic that wouldn't exist, or at least would be much less serious, without the strong forms of male sexuality. (Incidentally, it's essential that men refrain from sex until their circumcision wounds are completely healed, or they will actually increase their partners' risk.)

Gates writes, too, about the quest for a vaccine and treatment for prevention—early treatment that reduces the risk to others. This works especially well in preventing women from passing it on to their unborn children. But treatment—the contrarian Elizabeth Pisani again—however much it is the right thing to do, will not suffice for prevention. That's because the decades-long maintenance therapy has inevitable interruptions: people run out of medicine and can't afford the next refill, or they feel good so they skip their meds for a while, or they become resistant to the ones they've been taking. They feel good, they look good, they party, but they may not have a low enough viral load to prevent transmission.

The Gates Foundation is also active in family planning; some 200 million women around the world say that they don't want to have a baby in the next two years *and* that they don't use contraception. Gates funds implants and injectables and wants to develop forms of those that women can give themselves. You can see what this has in common with male circumcision: you make one decision, do one thing, and get long-term protection without further use of willpower. For women, this is particularly important, since their problem is often not their own willpower but men's will imposed on them. *If he won't use a condom, at least let him be circumcised.* Or: *At least I have my implant, so I don't have to beg him to wait while I put in a diaphragm or a sponge.* Protection from unwanted pregnancy and infection should not depend on men's goodwill or on a woman's ability to resist. If sex gets rough, vaginal tears get more common, and germs run rampant. If we can't protect a woman completely, we can at least improve her odds.

There is another epidemic that overwhelmingly targets women. Two notorious, brutal gang rapes occurred in India within a year—one in December 2012 in Delhi, one in August 2013 in Mumbai. In the first, six young men raped and murdered a student on a bus; they beat her male companion senseless with an iron rod and used it to penetrate her in such a way as to severely damage her intestines, which were surgically removed in a vain attempt to save her. The youngest rapist, a teenager, was sentenced by a fast-track court to the maximum juvenile punishment of three years; a second hanged himself in prison. The other four were sentenced to death, which is reserved by the Indian Supreme Court for "the rarest of the rare" cases; in March 2014, the Delhi high court upheld the death sentences. In the Mumbai case, five men attacked a twenty-three-year-old photojournalist; after tying up her male companion, they took her aside and serially raped her. She spent a month in the hospital. Five suspects were arrested and tried; three, who were repeat rape offenders, were sentenced to death in April 2014.

Together with my wife, a psychologist who works with young girls at risk for commercial sexual exploitation, I spent a month in India between these two events. Everyone we met told us that the December gang rape in Delhi had provoked the conscience of the nation. People from Mumbai said that this was the sort of thing that happened in Delhi but not in their city. When we traveled between Orchha and Agra (the site of the Taj Mahal), I did not know what had happened to a Swiss couple touring by bicycle between those cities in March: the man was tied up and made to watch while eight men serially raped his thirty-nine-year-old wife. Six were convicted and sentenced to life in prison.

These rapes have something of the character of what used to happen in the era of warring tribes and empires. The attacks are partly about sex but mainly about humiliation and violence. The perpetrators beat and immobilized the men and made them watch as the women they should have protected were savagely gang-raped. The men's humiliation was part of the script.

But most rape in the world is not like this. Most of it is done by a partner or other acquaintance, not by strangers. All of it restricts women's freedom by intimidation. Rape and other violence against women, including wife beating, "honor" killings, and other family assaults, are part of the structural violence women face throughout the world. In a 2012 WHO ten-country study of women fifteen to forty-nine years old, the number that had been forced to have sex ranged from 15 percent in Japan to 71 percent in Ethiopia. The percentage of women whose *first* experience of sex was rape—a foundation for lifelong fear—was 17 percent in rural Tanzania, 24 percent in rural Peru, and 30 percent in rural Bangladesh.

Social epidemiologist Karen Devries and her colleagues of the London School of Tropical Hygiene further analyzed the data in a 2013 paper. They found that 30 percent of women aged fifteen and older have experienced physical and/or sexual intimate partner violence (IPV). It is everywhere. The lowest regional averages are East

Asia (16 percent) and Western Europe (19 percent). East Africa and Andean South America are very high (around 40 percent), but war-torn central Africa is in a class by itself at 66 percent. South Africa, though, has a model program that cut rural IPV by more than half in two years.

Girls' education and, for both sexes, microfinance loans and school-based prevention programs work in various settings, and legislation to punish rapists and *not* victims has been adopted in many countries. India passed a new national law this year to increase prosecutions and severe punishments for rapists, including fast-tracking trials. However, it remains legal there for a husband to rape his wife, and during the 2014 Indian elections, the head of a recognized party in the state of Uttar Pradesh responded to the gang-rape convictions by saying, "Boys make mistakes. They should not hang for this. We will revoke the anti-rape laws." An associate and sitting state legislator, asked a few days later about the party head's remarks, went further: "Any woman, whether married or unmarried, who goes along with a man, with or without her consent, should be hanged. Both should be hanged."

Perhaps this outlook on hanging women for being raped was in the minds of the men who in May 2014 gang-raped two girls, fourteen and sixteen, near a village in the northern state of Uttar Pradesh, because after the rape they hanged the girls from a mango tree using the girls' own scarves. The abduction of the girls was witnessed; three brothers were arrested and confessed to the rape but not the murder, and two policemen were charged with complicity. Like hundreds of millions of rural people throughout the world, the girls did not have access to a toilet, so they were in the fields relieving themselves when they were attacked. The chief minister of the state, pressed by journalists, said, "You're not facing any danger, are you? Then why are you worried? What's it to you?" A home minister in a neighboring state said that rape is "a social crime which depends on men and women. Sometimes it's right, sometimes it's wrong."

The UN estimates that 600 million women today live in countries where intimate partner violence is not even nominally a crime. But lest it be thought that things are better in the United States than India, we should note the U.S. led the world in reported rapes in 2010 (about 85,000), while India, with four times the population, was a distant third, with only around 22,000 reported. However, experts say that almost one in two rapes gets reported in the United States, while only one in ten does in India. In addition, marital rape is not included in the statistics in India, while it is in the States. So given all the problems with getting accurate statistics, it is probably fair to say that the rape rates are not so different in the two countries.

Nor is the dismissive sort of official attitude we saw in India limited to faraway places with strange-sounding names. Consider that the U.S. Department of Justice was in 2014 investigating prosecutors in Missoula, Montana, for reportedly saying to an alleged rape victim, "All you want is revenge" and telling the mother of a five-year-old raped by a teen, "Boys will be boys." In 2012, in Steubenville, Ohio, the rape of a drunk, unconscious sixteen-year-old girl caused a scandal when her rapists used social media to brag about their crime and mock their victim. Because the two rapists were football stars on the local high school team, football fanaticism combined with what has been called a "rape culture" in the United States led to public defense of the criminals by town and school officials and to a partial cover-up and lies about what those officials knew and when they knew it. The coach testified as a character witness for the boys who'd committed the rape, said he was proud of them and saw no reason to bench them, and in one episode shouted expletives at journalists and threatened them and their families.

Little wonder that a worldwide response to such things has been mounted. Eve Ensler, author of *The Vagina Monologues,* began the V-Day movement—for violence, vagina, and Valentine's Day—in 1998. On February 14, 2013, Valentine's Day in the United States, V-Day launched a campaign called One Billion Rising, since almost one in three women, or about a billion, are raped or assaulted in their lifetimes. The goal was for people of both sexes to call for an end to such violence by assembling, striking, demonstrating, and dancing. Tens of thousands of people did demonstrate on that day throughout the world, and by all reports the V-Day event in 2014 was at least equally widespread and successful. Peaceful demonstrations rolled out for a total of forty-eight hours across the planet's time zones. UN Secretary-General Ban Ki-moon was quoted as saying, "The global pandemic of violence against women and girls thrives in a culture of discrimination and impunity. We must speak out. . . . I am proud to emphatically raise my voice and join the chorus of all those taking part in the One Billion Rising campaign." The goal now is to hold the worldwide event on February 14 every year.

Another issue targeted by One Billion Rising, as well as by WHO, Care International, the Carter Center, and other organizations, is female genital cutting (FGC). This name (unlike female genital mutilation) does not immediately alienate the people, mainly women, who perpetuate the practice but also does not minimize it as "female circumcision." It is much more invasive than male circumcision, involving at least the cutting out of part or all of the clitoris, the center of most women's sexual experience and the only human organ that has no purpose except pleasure. Often it also involves cutting away parts of the labia and even bringing the cut edges together to seal off most of the vaginal opening. As traditionally performed, it commonly leads to medical and surgical complications that can be life-threatening. It leads to a lifelong reduction in the quality of a woman's sexual enjoyment, which is often part of the goal.

In Africa (including North Africa) an estimated 100 million girls and women over the age of ten have undergone some form of this, with another 40 million in Asia and immigrant groups in the West. According to a 2012 report by Orchid Project, a charity focused on FGC, it remains legal in the United Arab Emirates, Yemen, India, and Pakistan and common although illegal in Malaysia, Indonesia, and Iraqi Kurdistan. A new American phenomenon, "vacation cutting," involves girls in African immigrant families being sent to Africa for the summer and being operated on there, often without advance notice and against their will. (Federal law banned the practice in the United States in 1996, and in 2013 it became illegal to transport girls for this purpose.) Although it is common in Muslim countries, it has no basis in Islam; a meeting of Islamic scholars in Cairo in 2006 resulted in a fatwa issued against FGC by the grand mufti of Egypt. The Carter Center is working to build an international religious consensus against it, as described in the former president's 2014 book.

Eliminating it is one goal of Molly Melching's organization Tostan—"breakthrough" in the Wolof language of Senegal. Melching moved there in 1974 and began building trust and friendships. Tostan's Community Empowerment Program, now deployed in eight countries and reaching 200,000 people, takes three years to implement in a village, with a trained facilitator fluent in the local language. In a June 2013 interview by Melinda Gates herself, Melching said it is "understandable that people are outraged when made aware of some harmful practices . . . but I learned in Senegal that aggressiveness does not lead to effectiveness . . . that combative, judgmental methods actually led to people shutting down and turning off."

She also explained, "We never accuse or fight the men, particularly the traditional and religious leaders. Rather we see them as allies, as important leaders for change in the movement. . . . There are many male heroes in this movement in the eight African countries where we have been working." The Tostan workers learned not to focus on one change and, especially, not to preach; Melching is justly proud of Tostan's inroads against child marriage and in favor of educating girls. But she also says, "The courage of the rural women in this movement to end FGC has particularly amazed and deeply touched me." In many cultures you are not considered a real woman unless you are cut, and if you are to avoid it as a girl, your mother, who was herself cut, must defy *her* mother and other women in her family and the village—not to mention the families of men who may reject you as a future wife.

Change can occur only village by village, and it works best if people end up thinking it was their idea. When Melching heard rural Senegalese women and men say that child marriage and genital cutting did not fit *their* values, she knew she was getting somewhere, and the fact that it took decades was of the essence. Now, about 5,500 communities in Senegal and 1,000 more in eight other countries have declared publicly that they have abandoned FGC. We don't have to imagine 100 percent follow-through to believe that Tostan made a difference. As a Wolof proverb says, "It is better to find a way out than to stand and scream at the forest."

CARE (Cooperative for Assistance and Relief Everywhere) is a large, old charity—I remember CARE Packages, which began in the late 1940s as U.S. Army surplus meals ready to eat, sent to pockets of hunger in postwar Europe. CARE has long been focused on world poverty but has now made bettering women's lives its strategic path to that goal. According to its statistics, 70 percent of the world's 1.3 billion poorest people are women; within that 1.3 billion, women contribute two-thirds of the working hours but earn 10 percent of the income and hold 1 percent of the property. Two-thirds of illiterate adults are women, and 60 percent of children not in school are girls; in India, make that 80 percent.

A wide-ranging report titled "Women's Empowerment" says, "A key breakthrough in CARE's evolving understanding of the underlying causes of poverty has been the explicit recognition of power as the currency of material and social well-being." This realization—I, for one, was surprised by the use of the language of the counterculture to reorient a large, old, conventional charity—led to the I Am Powerful campaign, designed (under CEO Helene Gayle) to "reposition" the organization's "brand." In 2008 a case study at Harvard Business School and a special issue of *Advertising & Society Review* were devoted to this rebranding, which worked like a charm. Once associated chiefly with emergency packages—and it does still respond to crises—CARE is now known as the largest organization using women's empowerment as a lever to lift the poor, men and women alike, out of poverty.

CARE's field research in India, Ecuador, Bangladesh, and Yemen shows that women feel empowered by self-worth and dignity, bodily integrity and freedom from coercion, control and influence over resources, and collective effort and solidarity. This may sound like a crisp summary of UN Women's *Transformative Stand-Alone Goal,* except that CARE was years ahead of the UN in making women's lives the top priority and no doubt served as a model. This starts with educating girls.

Again: girls who go to school postpone motherhood, have fewer, healthier, better-educated children, ensure better health for themselves *and* their husbands, are less likely to be abused by men, and earn more over their lifetimes. The impact is "dose"-related, improving with each added year of school, with a transgenerational multiplier, as educated girls have more educated daughters and yet more educated granddaughters.

In a project in northern Benin, girls' school enrollment doubled in four years, an increase of eight thousand girls, with plummeting dropout rates and rising academic achievement. In southern Sudan, after years of genocidal war, CARE established the first real school system; where most people had never met an educated woman, they were introduced to their own educated fellow countrywomen, who were sent out to visit remote villages and enroll girls. CARE developed a series of schoolbooks, based on many interviews with southern Sudanese girls, in which characters "are smart, engaging and bold girls who challenge local notions of what a girl 'should' be." CARE also funds centers to shield women from violence within refugee camps in Kenya and Chad, as well as on Sri Lanka's east coast in the wake of decades of civil war.

In several places around the world, CARE has adopted versions of microfinance, developed in the 1970s by Muhammad Yunus and the Grameen Bank (among others) and laureled with a Nobel Peace Prize in 2006. Many banks and NGOs have applied it over the years; the consensus was that microloans work better if given to women rather than men. Grameen Bank clients have been more than 90 percent poor women, usually in groups (solidarity lending). This was based on the idea that women are more likely than men to repay, to put the loan to good use, and to respond to social pressures from other women in their groups.

These assumptions have been confirmed in some studies but not in others, and they remain controversial. That is why CARE has implemented its "Microfinance-Plus" program, which acknowledges that the loan may not be enough. In India, for example, the "plus" involves training in human rights, health, and governance. A controlled study showed that women who had social, political, and business-development training for three years developed greater independence and increased influence and equality in the home. They spent 125 percent more money on the education of their children and 43 percent more on health care than women in a control group.

The research also raised concerns about the much more widespread microfinance-only programs; these can cause reactive abuse of women by men, who may also use their wives to get money. But in Niger, 150,000 women have been recruited to Women on the Move, which emphasizes small savings-and-loan co-ops, with similar positive results—including a small but unprecedented number of women running for and winning local office.

Finally, as part of its fight against AIDS, CARE has for two decades helped sex workers in Bangladesh. Through a local organization, SHAKTI (Stopping HIV/AIDS Through Knowledge and Training Initiatives; the acronym is also the Bangla word for "strength"), CARE workers have trained women to negotiate safe sex and condom use and to form self-help groups. One named itself Durjoy Nari Shanga—

"Difficult-to-Conquer Women." In the first phase, condom use went from 12 percent to 73 percent for brothel-based sex workers and 51 percent for those working the streets. Harassed by police, they learned to say, "Like you, we have an equal right to an income." Looking back, they said, "We realized we are also human beings." It's not your grandmother's CARE.

We say that abuse of women leads to a kind of slavery, but of course there is also plenty of outright slavery, no metaphors needed. While for obvious reasons it is difficult to get accurate numbers, good sources say that there are between 21 and 30 million slaves in the world today, *more than at any time in history.* According to the UN's Global Report on Trafficking in Persons for 2012, about half the trafficked people detected in one year were bought and sold for sexual exploitation. According to the U.S. State Department, some 800,000 people are trafficked across international borders each year, around 80 percent of these are women, and around 70 percent of those are sexual slaves. This means that more than half (nearly 450,000) of the people bought and sold each year are girls and women bound in sexual slavery.

Nonsexual slave labor includes forced begging (for the slave owner's benefit), carpet weaving, brick making, wood burning for industrial charcoal, mining and quarrying, commercial fishing, and coffee and chocolate growing, among many other harsh jobs. Almost everyone reading (or writing) this book buys products every day that are partly produced by slaves.

Women and girls make up an estimated 55 percent of such labor slaves. However, the most profitable way by far to make use of someone you have bought and paid for is to prostitute her (or, in much smaller numbers, him). It is the third most profitable illegal business in the world, after drugs and weapons trafficking, totaling over $30 billion a year internationally, almost half in *industrial and postindustrial* countries. Sexual slaves are overwhelmingly female. Most of them are under age eighteen when sold, and resistance is beaten out of them early on. Severe beatings result from escape attempts or disloyalty to your owners, such as talking to people trying to help you. You service large numbers of customers, who also beat and otherwise abuse you; welts, bruises, and cigarette burns are common. Sexually transmitted infections are rampant. Many sex slaves are abused until they die—often in just a few years.

Opposition to girls' education converged on the risk of slavery in Nigeria in April 2014, when the violent Islamist group Boko Haram—the name means "Western education is a sin"—kidnapped more than two hundred girls from their school in a northern village. A few weeks later, the organization's leader threatened to sell the girls, saying he would "give their hands in marriage because they are our slaves. We would marry them out at the age of 9. We would marry them out at the age of 12." Forced child marriage is probably the best fate they would face if they were actually sold by their captors. A wave of anti-government protest, rare in Nigeria, demanded action. By June the girls' location was approximately known, but rescue was considered impossible; the group could be expected to murder the girls during any attempt, since these extremists have slaughtered hundreds of people in the past, children included. They have attacked the girls' village and killed some of their parents. In September, the girls were still captives, some dispersed to unknown places, but the Nigerian government was negotiating for their possible release.

Given the power of girls' education, not surprisingly some men oppose it with all the violence of a male supremacist past. The Pakistani Taliban burned down schools and tried to murder a 15-year-old who had raised her voice in favor of education, shooting her in the head; she barely survived, but two years later, she is loved by the world as "Malala," and her voice is stronger than ever. Ten men who conspired to murder her were arrested in September 2014, and in October she won the Nobel Prize.

And in other echoes of the darkest days of the human past, the Islamist terror army known as ISIS (or ISIL, or the Islamic State) has mounted genocidal campaigns that include the mass murder of men of other religious groups and the wholesale rape and enslavement of those victims' wives and daughters. For example, between 1,500 and 4,000 women and girls from the Yazidi minority group in Iraq have been abducted, raped, and tortured, some sold to terrorist fighters as brides, others killing themselves to avoid sexual slavery.

But others have been and are being saved, throughout the world. The CNN Freedom Project collaborates with Free the Slaves, which exposes all forms of slavery, mounts raids to free child slaves, pays for their schooling and books afterward, helps families start businesses so that they won't sell their children, and urges national governments as well as the UN to take more concerted action to root out and prosecute slave owners and traffickers. Boycotts of companies identified as using slave labor, long prison terms for violators, rescue operations, campaigns directed at customers, and other strategies are growing in size and effectiveness, but much more action is needed. Free2Work, a program of the organization Not for Sale, grades companies in slave-using industries. For example, 40 percent of all commercial chocolate is produced with slaves playing a key role. All the most recognizable names on the wrappers of the chocolate we buy earn grades of C- or worse, mostly D-.

The Coalition Against Trafficking in Women coordinates activities of networks fighting sex slavery, in particular, throughout the world. The International Princess Project is devoted to restoring the dignity of women formerly enslaved in prostitution in India, by advocating for them, giving them support to heal, and starting sewing centers around the country that train them and enable them to work as free women. They make pajamas under the brand name Punjammies, and they get the profits.

But don't imagine this is someone else's problem. If you live in any American city, you have sexual slavery in your backyard. The United States is a major buyer of girls and women, and there are hundreds of thousands of our own homegrown underage girls forced into prostitution in this country. Properly known as CSEC—commercial sexual exploitation of children—it is by definition involuntary, since a child cannot consent. These girls turn over their money to a man or woman who runs their lives and may, as well, formally own them. They usually have no family to go home to or came from one that is brutally abusive.

In Atlanta, where I live, the problem is bad because the airport is an international hub surrounded by neighborhoods where the most vulnerable children live. The average victim is an African-American girl between twelve and fourteen. Most of the customers are white men from the Atlanta suburbs, but many come from New York, Chicago, and other cities. They can purchase a package including round-trip airfare, pickup at the airport, and sex with a child and return home to their families in time for dinner. Some men who have been arrested had infant seats in their cars.

Ann Kruger (my wife), Joel Meyers, and their students at the Georgia State University Center for School Safety implemented a program in 2007 called Project PREVENT (Promoting Respect, Enhancing Value, Establishing New Trust). Using socioeconomic maps to locate schools in the most disadvantaged areas, they train group leaders to conduct after-school intervention sessions with sixth-, seventh-, and eighth-grade girls who have not been subjected to commercial sexual exploitation but—because of the neighborhoods they live in, their families and schools, and what they already know—are at high risk. The researchers do not explicitly ask about sex, but the girls spontaneously speak about it.

One said, "It's a whole bunch of girls at Metropolitan [Avenue]. I will be at the Chevron on Metropolitan, and they will knock on people's windows." Another said, "They will walk up to the prettiest cars. They will think, 'if they can afford them, they can afford me.' Sometimes the car will take them away. The security guards don't say nothing. They young. Like twenties." Another said, "These girls offer something strange for a piece of change." They talked explicitly about the "boyfriend-to-pimp" transition, one saying, "Some guys you date will try to get you to do things and say they will never do this and that to you and before you know it you're in the back of a car."

What they say about their own lives is chilling. Older men lurk on the periphery of school grounds and teen recreation facilities. One girl was approached by a strange man who said, "If I were your pimp, I'd let you wear lots of makeup." Girls used sex-industry terms to describe what happens at the Metro, a skating rink. One said that the main purpose of going there was to give boys lap dances and that if she missed a week she might lose her "regular customers." Another described how a boy will lie on his back and a girl will "do a split" on him. They openly discussed the merits of stripping as a career, and one said, "Whatever y'all do in the adult clubs is innocent compared to what happens at Metro." There is little to no adult supervision, lighting is low, and sexual activities predominate. When the girls leave, at three A.M., there are adult men waiting to prey on them.

Georgia Care Connection estimates that 7,200 men each month pay for sex with underage girls statewide, exploiting an average of 100 girls per night. At least 400 different girls are trafficked each month. When Kruger and Meyers began their project, there was one home for girls rescued from sexual slavery: Angela's House, one of three like it in the nation; it had six beds. Johns and pimps were rarely arrested, but girls frequently were; they were either put in juvenile detention or released back onto the street. In a moving press conference in 2007, then-mayor Shirley Franklin revealed that a friend's father had abused her as a girl. She began the "Dear John" campaign to pursue and discourage customers. *Not in my city,* she said, *not on my watch.*

Since 2009, it is mandatory to report knowledge of sexual exploitation as child abuse. In 2011, a new law made it easier to prosecute human traffickers, impose harsh penalties, and protect victims. Detained girls are no longer criminals. The state has training programs for law enforcement personnel, prosecutors, mental health workers, and medical personnel, as well as a "CSEC 101" workshop for the general public. Prosecutions and convictions of johns and pimps are up, and prison terms are longer. In 2013, a new law required posting of the telephone number of the National Human Trafficking Hotline in places where girls are at risk.

Shared Hope International is a Christian abolitionist group that has a Protected Innocence Challenge. It rates states based on "41 key legislative components that must be addressed in a state's laws in order to effectively respond to the crime of domestic minor sex trafficking." Georgia's grade was raised from 75 (C) in 2011 to 82 (B) in 2013, ranking 11th in the anti-CSEC fight. So if you were feeling holier-than-thou compared to Georgia as you read about this, I suggest you check whether your state is one of the ten ahead of us or one of the thirty-eight behind us (one was tied with Georgia). You might be surprised. California, for instance, got a 50 (F); New York got a 66.5 (D); Ohio scored 72.5 (C); Massachusetts, though, was just behind us, at 81. A's went only to Washington State, Tennessee, and Louisiana; Texas was next best, with 87.5. In 2013 California recognized Georgia as a leader and a model for statewide action.

Georgia, at least, is getting better every year. In January 2013, a state trooper stopped a car for speeding and, because he had been trained to see it, recognized that the adult man driving was holding the passenger, a young girl, against her will. The *Atlanta Journal-Constitution* reported, "The deputy separated the young girl from the driver, and she told him that she was being sold for sex. The girl was rescued. She told GBI agents later that she had prayed to be rescued, and she felt God had answered her prayers."

All these legal efforts are important, but at least equally important is the effort to raise the self-esteem of girls at risk and to restore it to those who have been exposed to this grievous harm. This is an uphill fight against a hugely lucrative criminal enterprise taking advantage of men's lust and girls' vulnerability, but in the last decade new recruits and new weapons have made it a fight we can win. New digital technologies have also been deployed; girls who can't call the hotline can text to "BeFree."

It may seem, after these crimes against women and girls, a luxury to talk about their roles in STEM fields—science, technology, engineering, and mathematics—but this is critical going forward; issues relating to self-esteem are holding girls and women back, and that is a general feature of situations where women lag behind men.

In the 1990s, social psychologist Claude Steele launched a new field of research by showing how easy it is to worsen the test scores of African-Americans just by activating, however obliquely, their belief that blacks don't test as well as whites. You didn't need to say anything about testing, just highlight racial awareness or change the race balance of the classroom. It is called stereotype threat, and it turns out that these same processes apply to women. A 2008 analysis by psychologists Hannah-Hanh Nguyen and Ann Marie Ryan found confirmation in many studies, but interestingly, subtle cues, like giving a pretest questionnaire, hurt women's performance more than explicit ones, like stating that men do better on the test.

Today, studies probe subtle factors that increase or decrease threat, and test interventions to reduce it. Analysis of seventeen years of research in 2013 by educational psychologist Katherine Picho and her colleagues showed that the impact of stereotype threat is greater in countries with less gender equality and that it declines between high school and college. Sociologist Andrew Penner had already shown that countries differ greatly in the size of the gender gap in high performance; for example, the Czech Republic and Norway have about three times the gender gap of the United States at the ninetieth percentile of a commonly used math test. This national difference cannot have a biological explanation.

An ingenious 2013 experiment by psychologist Emily Shaffer and her colleagues revealed that framing has a major impact. College students were given one of three bogus news articles to read before taking a math test. In one (the Balanced Condition), the article was called "Recent Study Shows Men and Women Nearly Equal in STEM"; an expert was quoted as saying that women had gained a lot of ground and were just as able as men in STEM. She made reference to a Fields Medal in mathematics supposedly won by a woman—no woman had then won this "Nobel Prize of math," although one did in 2014—and other positive statistics. In the second (Unbalanced Condition), the article was the same *except* the expert went on to say, "However, it is important to realize that women haven't yet achieved equality to men. I hope that a future study reveals even more progress made by women in STEM." This article ended with "more strides need to be made." The third article (Control Condition) was just a brief history of the college.

Women in the Balanced Condition did better relative to men than women in the Control Condition, and better than women did in the Unbalanced Condition, while men showed no differences. Yet both the Balanced and the Unbalanced Conditions called attention to gender differences, which usually hurts women's performance. What was new here was the framing: "Calling attention to women's success and subsequent equal representation in STEM was clearly beneficial for women without harming the math performance of men." This suggests an intervention strategy.

Research around the world (that is, around our species) shows what works. In a 2013 study by cognitive psychologist Annique Smeding and her colleagues, French middle school girls did worse than boys on a math test but not on a verbal test if the math was given first, while there was no sex difference on either test if the *verbal* was given first. Psychologists Melanie Steffens and Petra Jelenec studied ninth graders and college students in Germany in 2011 and found that implicit stereotypes held women back in math even when they were not aware of them, suggesting that addressing implicit beliefs will help. And a 2012 study in Uganda showed that women in single-sex schools were not affected by stereotype threat while women in coed schools were.

The impact of single-sex schools remains controversial; there have been many studies, but they don't all find a difference. There is an argument to be made that they are better for girls *and* boys, which is surprising given that not long ago "separate but equal" was considered the ultimate hypocrisy. Now that girls do better than boys in school, however, proponents argue that boys can't thrive in the female-forward environment of coed schools. Yet some of those who are looking out for girls' interests see an advantage for them, too: in an all-girls school, every role model, every success story, be it in elections, sports, science, or math, is female.

Israel offers a unique take in two studies in different populations. Educational sociologist Hanna Ayalon published a paper in 2002 called "Mathematics and Sciences Course Taking Among Arab Students in Israel: A Case of Unexpected Gender Equality." It is clear that Israeli Arabs, although citizens, are not equal citizens. But ironically, because course offerings in Israeli-Arab high schools are thin in social sciences and humanities, girls take almost as many math courses as boys, and almost as many as Jewish *boys* do in *their* schools. So in math and computer science, Jewish girls lag behind Arab girls and all boys.

Except for one group of Jewish girls. Sociologist and anthropologist Yariv Feniger, studying more than twenty thousand Israeli high school students in 2010, found that girls at all-female religious schools, while they did not enroll very much in math, physics, or biology, were far more likely to take computer science than girls in coed schools. Again, it was a question of opportunity. Late in high school, boys in same-sex religious schools were deepening their religious studies, while families were urging girls to learn computer science. A friend of mine, a nonreligious Israeli-American with two Ph.D.s (one in physics) and a career in computers, insisted that her American daughter (now an architect) learn how to program. "My mother's mother," she said to the young woman, "told my mother that if she learned to sew she would never go hungry. I'm telling you this for the same reason."

Whatever the reason, girls should be taking computer science, physics, math, engineering, everything. And increasingly, they are. A 2013 book by Amy Sue Bix—*Girls Coming to Tech!*—documents the history of women at the three top institutes of technology in the United States: Caltech, MIT, and Georgia Tech. Just before World War II, around 1 percent of the students at these schools were women. In 2013, 45 percent of the undergraduates and 31 percent of the graduate students at MIT were women. Bix writes, "Before 1952, Georgia Tech had no undergraduate women; in 2011, Georgia Tech led the nation in granting engineering bachelor's degrees to women. In early 1970, Caltech had zero female undergraduates; in 2011–2012, women comprised 39 percent of undergraduate enrollment."

Nationwide, most engineering schools lag behind these three trendsetters, but other programs are following their lead. The more that do, the more models there are for the next generation, and the next. The nonprofit organization Code.org has in about a year reached thousands of K–12 teachers and millions of their students with programs like An Hour of Code. Both sexes are eligible, but the nonprofit says it has reached more girls in the past year than have learned to code in all of previous history. With the support of tech industry leaders, Code.org mounts teacher-training programs to keep up students' interest, which has been keen, and offer many online learning experiences.

Larry Summers, when president of Harvard, publicly said that because men are more variable, women might not have the same abilities in math and science as the best men—a statement that would barely have been noticed a few decades earlier. Given the proven impact of stereotype threat, his claim probably diminished the math and science performance of girls and women at his and other universities. He was led to resign (for this among other reasons) and replaced by Harvard's first woman president. In the fall of 2013, under pressure, Summers withdrew his name from consideration to become the next chairman of the Federal Reserve, probably the most important banking position in the world. Janet Yellen, his main rival, in 2014 became the first woman in that role. Her word, not his, is moving markets worldwide; her judgment will stabilize economies and determine the prosperity of millions.

An elegant 2013 study by psychologists Katie Van Loo and Robert Rydell showed that priming women with a sense of their own power protected them from stereotype threat effects. This reminded me of a cartoon. Two nervous men in suits are sitting in the waiting room of a fancy office. One confides to the other, "They say if she tugs her earring, you're done for." That was in the early days of women in the executive suite, and it was funny because it was surprising. Now women run technical companies that employ thousands of engineers, as well as countless nontechnical ones that also employ technicians. The women on top understand more than enough of what the geeks have to say to them, but they get to the top because they also understand their customers and have a deft, nonconfrontational managerial style that many men can't imitate. Women in business, as we'll see, are better at keeping their egos and anger out of it, and that often makes for better judgment and better leadership.

## *Chapter 10*

✦

# Billions Rising

But, you fairly object, power is power, and it corrupts, regardless of its chromosomes. I can't disagree entirely, but I do disagree mostly. So how do I think the world will be different in the female-forward future? Well, for one thing, sexual abuse of either sex by people in power—something that dates back to the dawn of power itself—will decline by almost the same extent that women replace men.

As I began work on this book, the U.S. Senate was probing the recent antics of thirteen Secret Service men. While in Cartagena, Colombia, in advance of a visit by the president, they hired prostitutes, risking the security of the man they had sworn to take a bullet for, sullying the agency, and hurting their families.

At the same time, a former senator from North Carolina, almost elected vice president in 2004, awaited a verdict on the charge that he used campaign funds to support a child he'd fathered with an aide, whose parentage he'd tried to pin on another aide. All this became public while his devoted wife was dying of breast cancer; she responded with hysterical shame and grief. The only question in the case was whether donated funds were campaign contributions; the jury was deadlocked due to lack of absolute proof that they were, and a mistrial was declared.

Just a month earlier France had chosen a new leader; a different man would almost certainly have become president of that country, had he not been accused a year before of sexually forcing himself on a New York City hotel maid, only the latest in a long series of similar allegations against him. At the time he held one of the world's most important economic posts, in which he was replaced by the first woman ever to hold it: the brilliant and elegant Christine Lagarde, now head of the International Monetary Fund.

As I neared the end of my work on the book, the mayor of San Diego, previously in Congress for twenty years, was forced to resign from office and later sentenced to three months of house arrest and three years of probation. Nineteen women had accused him of sexual harassment while in one office or the other, although he pled guilty to only three assault charges, all while mayor: forcibly restraining a woman at a fund-raiser (a felony), kissing a woman on the lips without her consent, and grabbing a woman's buttocks while posing with her for a photo op. The plea bargain included apologies, but when he resigned a few months earlier, he had said that he was the victim of "lynch mob" hysteria. He was released from house arrest on April 7, 2014, saying he hopes to regain his integrity.

Meanwhile, the embattled mayor of Toronto, who months earlier had admitted to smoking crack cocaine—his excuse was that he was in a drunken stupor at the time, so he didn't know what he was doing—stood accused by staff members of having used drugs with a prostitute while celebrating St. Patrick's Day—as mayor, of course. He blamed his enemies for persecuting him, and in April, 2014, he announced his run for reelection.

However, in September, trailing in the polls and diagnosed with cancer, he withdrew from the mayoral race and announced a run for City Council. He is swapping campaigns with his brother Doug, who will pick up the dropped mayoral race as Rob runs for Doug's accordingly vacated Council seat.

These are only the most recent examples. Going back a little further: the governor of California, a famous actor who married into America's most elite political family, fathered a child by his children's nanny, who lived in his home for a decade without his wife's knowledge of their connection. A sitting governor of New York, then said to have presidential prospects, hired a prostitute linked to a crime syndicate he himself had pursued; he soon resigned, and eventually his marriage ended. A sitting New Jersey governor was forced from office by his admission (preempting outing) that he'd had an affair with a man who had been his security aide and who was now planning to sue him for harassment.

A married, deeply religious South Carolina governor disappeared and was out of touch for five days, claiming to have been hiking the Appalachian Trail; he was instead in Buenos Aires with an e-pal he had fallen in love with. "Hiking the Appalachian Trail" became a euphemism for adultery, but he was returned to Congress in 2013. A married, sitting U.S. congressman texted half-naked photos of himself and explicit sexual messages to several women from his cell phone, lied about it, and then resigned from office. Two years later, in 2013, he ran for mayor of New York, claiming successful therapy, remorse, and rehabilitation, but it turned out he had repeated his former offenses a year after his supposed rehabilitation. As has often happened in the past, some public men cannot control their behavior even under intense media scrutiny.

And, of course, a president of the United States got oral sex in the Oval Office from a twenty-two-year-old intern bringing him pizza; this, and his lies about it, resulted in his impeachment, greatly damaged his last years in office, and probably prevented his vice president from succeeding him, with momentous consequences for the economy, social policy, and war.

Then, too, there was the all-male, theoretically celibate hierarchy of the Roman Catholic Church, which for decades (at least) perpetrated, permitted, and covered up systematic child sexual abuse throughout the world. Child molesters in New York's Orthodox Jewish community have also been in the news; we don't need to be told the gender of the perpetrators, including a rabbi and a religious school official, nor that of the leading rabbis in the community, most of whom joined the cover-up and shunned parents who dared to report their children's abuse to civil authorities. The trial of a football coach at a major university, accused of raping many boys over many years, some on his school's premises, resulted in the school paying $60 million to twenty-six victims and in a thirty- to sixty-year prison sentence for the perpetrator. And revelations about New York's Horace Mann School, one of the nation's most prestigious prep schools, brought victims out of the woodwork day by day, exposing another huge cover-up of sexual abuse; in this case, the statute of limitations prevented prosecution.

These are just a few examples. All have to do with sex. But there are other kinds of scandals, of course: a former governor of Illinois recently went to jail for fourteen years for corruption, arrogant and unrepentant throughout. (Incidentally, four of the last seven governors of that state, all male, have done prison time for corruption, fraud, bribery, racketeering, or related charges.) A Louisiana congressman is serving thirteen years after a similar conviction; he had $90,000 in ill-gotten cash stored in his freezer. Three years earlier, a four-decade New York congressman and hero of the African-American community was severely censured for corruption by the House Ethics Committee and lost his most important positions in Congress.

Women can be corrupt, too—many have been convicted of embezzling, for instance, and Martha Stewart went to jail for insider trading—but they do these things much less frequently, even in proportion to their much more limited access to opportunities for high-level crime. More importantly, women are not as apt to drift into the clash of collective egos that leads to war—and worse. At this writing, at The Hague, General Ratko Mladic, ex–army chief of the Bosnian Serbs, is on trial for murdering more than seven thousand boys and men in Srebenica in 1995, among other crimes; he has taunted survivors from the defendant's box in his genocide trial. Charles Taylor, former president of Liberia, was recently sentenced to fifty years in prison for crimes against humanity. And for six years Omar al-Bashir, still president of Sudan, has been under indictment for the genocide in Darfur. Others, like Pol Pot, who murdered millions in Cambodia and died of heart failure at seventy-three—although there were also rumors of suicide, since the Khmer Rouge was about to turn him in—and Slobodan Milosevic, who died of natural causes while on trial at The Hague, escaped punishment. For the genocide in Rwanda in 1994, when at least 800,000 people were murdered in one hundred days, the International Criminal Tribunal for Rwanda has indicted ninety-three people, of whom sixty-one have been convicted. And at this writing, armed men are ravaging the Central African Republic, raping and killing large and growing numbers of victims, as is the ISIS terrorist army, which as we have seen has made rape a routine part of its own genocidal campaign in Iraq and Syria.

These people come from many different cultures and religions on three continents, with varied histories, backgrounds, and political views, but almost none of them are women. Most are implicated at least indirectly not just in murder and maiming but also in rape. The International Criminal Tribunal for Rwanda was, it says, "the first institution to recognize rape as a means of perpetrating genocide." The International Criminal Court has also applied this to Sudan. Because throughout history rape has been an integral part of war and of the process of genocide—the attempt to physically eliminate a whole people—we can only welcome this recognition of one of the most important ways men have brutalized women. Rape in these situations is almost invariably a male crime, overwhelmingly committed against women, and some of the victims are raped in such a way as to destroy their reproductive capacity or murder them. In any aspect of genocide, women are seldom the criminals, often the victims; in the rare cases when they do participate in these crimes, they are almost always following men's lead.

Elizabeth Cady Stanton's great speech, the one that launched a quest for women's rights, predicted much about how the world would improve with increased power for women. Recall that every argument, she said, mounted to justify extending the vote to landless men, to new male immigrants, and finally to men who had been slaves a few years earlier must equally be seen to justify women's suffrage. And she did not stop there: "All these arguments we have to-day to offer for woman, and one, in addition, stronger than all besides, the difference in man and woman."

But she did not stop there either.

She said that the "aristocracy" of sex is the most hateful and unnatural, "invading . . . our homes, desecrating our family altars, dividing those whom God has joined together, exalting the son above the mother who bore him, and subjugating, everywhere, moral power to brute force."

She described government by men as social, civil, and religious disorder. "The male element is a destructive force, stern, selfish, aggrandizing, loving war, violence, conquest, acquisition, breeding in the material and moral world alike discord, disorder, disease, and death. See what a record of blood and cruelty the pages of history reveal!" She referred to the slavery and slaughter, "inquisitions and imprisonments, pains and persecutions, black codes and gloomy creeds" with which the human soul has had to contend over the centuries, "while mercy has veiled her face."

She said, "The male element has held high carnival" and "run riot from the beginning, overpowering the feminine element everywhere." She described society itself as a mere reflection of men, without any real influence by women, resulting in "the hard iron rule" that obtains in state, church, and home. There is no surprise, she argued, at the fragmentation and disorder of everything, "when we remember that man, who represents but half a complete being, with but half an idea on every subject, has . . . absolute control."

She said, "The strong, natural characteristics of womanhood are repressed and ignored" as dependent women strive to mimic men in order to gain any influence at all. "She must respect his statutes, though they strip her of every inalienable right. . . . She must believe his theology, though it pave the highways of hell with the skulls of newborn infants, and make God a monster of vengeance and hypocrisy." Women are forced to make the best of whatever world men offer them. "To mourn over the miseries of others, the poverty of the poor, their hardships in jails, prisons, asylums, the horrors of war, cruelty, and brutality in every form, all this would be mere sentimentalizing. To protest against the intrigue, bribery, and corruption of public life, to desire that her sons might follow some business that did not involve lying, cheating, and a hard, grinding selfishness, would be arrant nonsense."

And she pointed to the greatest irony of all, the fact that "now man himself stands appalled at the results of his own excesses, and mourns in bitterness that falsehood, selfishness and violence are the law of life. The need of this hour is not territory, gold mines, railroads, or specie payments, but a new evangel of womanhood."

Some people want a manned mission to Mars, but I think that can wait. What we need is a mission to earth, and fortunately it has already started.

Let's look at just a few of the more dramatic changes. In my mother's generation it was not considered safe for women to drive, although she boldly did. Now we know that women have far fewer accidents than men at all ages, a difference particularly pronounced in the youngest and oldest people. They are just plain better drivers.

When I was in high school, in the 1960s, girls' sports were almost nonexistent. By 1978 *Time* magazine had a cover article, "Women in Sports," describing an ongoing revolution. The education law known as Title IX was six years old. Authored by Congresswoman Patsy Mink and Senator Birch Bayh, it prohibited institutions receiving federal funds from discriminating on the basis of sex.

By 2014 the number of women in high school sports had increased elevenfold from pre–Title IX levels, in college more than twelvefold. Women have transformed Olympic sports as fans of both sexes fill stadiums to watch them compete. Since 1978 I've continued to show my students a slide of that old *Time* cover in my lectures on sex differences. It depicts a young woman playing lacrosse, her lovely features shaped in a typical higher-primate threat face, common to the innate wiring of expression in all monkeys, apes, and human beings. I recently added a photo of soccer star Mia Hamm playing her fierce game with that same expression on her face. Women compete, women fight, women want to win, but they rarely take it to the dangerous extremes men have pursued for so many centuries.

However, they are not just playing: 16 percent of students at West Point are now women, up from zero in 1975, and reflecting the current proportion of women in the armed forces. Although the proportion has been stable for decades, plans are in place to increase both. Women have been deployed to conflicts in Afghanistan and Iraq and are more often demanding and getting combat roles. They have often flown fighter jets and Apache attack helicopters in combat. Tammy Duckworth, originally from Bangkok, Thailand, and elected to the U.S. Congress in 2013, flew combat missions in Iraq. In 2004 she had lost both her legs and partial use of one arm when a rocket-propelled grenade hit her Black Hawk helicopter. In 2013 the Pentagon lifted its combat exclusion role for women in the infantry, opening hundreds of thousands of frontline positions. Women are scheduled for admission to elite Army Ranger training in 2015 and Navy SEAL training a year later.

Does this mean women will become just like men? I don't think so. They are and will continue to be good soldiers, but the fact that women want to serve does not negate the differences. As women become true partners of men, wielding influence at every level of society, the risk that anyone, male or female, will have to go to war will go down. This is the feminization that Steven Pinker writes about in *The Better Angels of Our Nature* and that in part explains the great long-term decline in violence. The growing participation of women in leadership roles is an unstoppable trend, and its consequences will be the ones Elizabeth Cady Stanton predicted: a more humane and safer world for everyone.

The evidence for the trend itself is clear. As we have seen, girls and women in the United States do better in school than boys and men. The gender gap has been almost completely reversed. The majority of young people now graduating from high school, entering college, and finishing college are female, and if males were not favored for purposes of diversity—in effect, many colleges now have de facto affirmative action for males—the imbalance would be worse, an outcome not necessarily welcome to women competing for men in the hookup culture. But it would inevitably mean more women leaders. Medical, law, and business schools continue to slightly favor men, but ongoing trends may equalize those numbers soon. One of the more interesting new psychological findings is that not only are leaders more likely to be the eldest child in their families, but this effect is larger for girls, yet another factor favoring women's future achievement.

In the twenty-first century, and especially since the Great Recession of 2008, men's rates of job loss and overall unemployment have been much higher than those of women. According to the National Women's Law Center, from the beginning of the recovery to July 2013, American women regained almost 95 percent of their lost jobs while men regained only 65 percent. In the same time period, women's unemployment declined from 7.6 to 6.5 percent and men's from 9.9 to 7.0 percent. Briefly in 2013, men's unemployment edged a tad below women's, but that was due to men leaving the workforce, not to men getting jobs.

We've seen positive trends in the leadership of major corporations, but the percentages of women at the top remain low. Not so for new businesses; the number of women going into business for themselves in the twenty-first century is one and a half times that of men. Between 1997 and 2014, the number of businesses owned or majority-owned by women topped nine million, about 30 percent of businesses, employing more than seven million other people. Given that women-owned businesses are increasing in number much faster than men's, women's share can only climb. As *Forbes* magazine put it in 2012, "Entrepreneurship is the new women's movement."

Similar trends are evident in the nonprofit sector. In September 2013, when Helene Gayle, the physician CEO of CARE, was asked on CNN for her thoughts on being the first woman and the first person of color to head that massive charity, she said, "I think it was just a matter of time. Women are playing all sorts of different roles in our society. . . . If it hadn't been me, it may have been another woman. . . . There will be one day when this is not going to be so unique." Women lead around 19 percent of America's four hundred largest charities, nearly four times the percentage of women running Fortune 500 companies.

But women's managerial role is changing at its roots. A front-page article in the *New York Times* of September 7, 2013, by Jodi Kantor described the then-latest graduating class at Harvard Business School, the first products of an experiment to change the school's culture. HBS offered assertiveness training for women (including teaching them how to raise their hands) to improve grades in the all-important case study sessions, as well as coaching about the outside-of-class culture, dominated by super-wealthy male students, to discourage women from trading sexual and social favors—or, worse, deliberately holding back on the success they could have in courses—in order to develop relationships with male stars.

This is not the college hookup culture, although it involves sex; these women are not nineteen but twenty-seven. They are no longer postponing real relationships; they are seeking them, and they will never again have access to this kind of pool of unmarried successful men. For decades women at HBS were openly denigrated by men, lacked role models, knew they would not have the same real-world opportunities, and retired into the background. But Drew Gilpin Faust, a distinguished Civil War historian and Harvard's first woman president—the one who replaced Larry Summers after his sexist gaffe, among other blunders—made up her mind to change this culture and in 2010 began to appoint administrators at HBS who would do that. The changes were strong and pervasive.

There was certainly pushback from some men, but even they conceded that the program was working and that they paid attention to new data showing that women investors do better than men and that companies with more women on their boards do better as well. Interviewed at a graduation party celebrating the fiftieth anniversary of women at HBS, one young couple about to get married joked about their status, because the woman had a job and the man did not. She said she would be supporting him, and he called her his "sugar momma." She said, "It's, like, perfect, I came to HBS just to find him, a stay-at-home dad, my trophy husband!" Kantor, interviewed on WNYC after her article came out, had heard from the fall semester's newly enrolled women. They were puzzled by the note of pessimism in her story, finding the school a very good place for women and a very positive experience.

Stanton did not just talk about fairness for women, or even just the improvements in society that would come from declines in bad male behavior. She envisioned positive changes that would follow from the rise of women. She alluded to man as "but half a complete being," and said sardonically, "To mourn over the miseries of others, the poverty of the poor, their hardships in jails, prisons, asylums, the horrors of war, cruelty, and brutality in every form, all this would be mere sentimentalizing."

Is it? Or is there something more than the negative—the *not-male,* as in refraining from violence and abusive sex—that we can legitimately expect from women as they gain more power? Evidence is growing that there is.

It is well established that there is a gender gap in political attitudes, and not everyone likes the result. Neither Bill Clinton nor Barack Obama could have been elected without disproportionate support from women, and the media frequently mention this as a problem for Republicans in the United States. This is also true in Britain, Norway, and a number of other advanced countries, but not all of them. Also, a simple left-right distinction will not explain all gender differences: Women are more in favor of government-sponsored social programs and more opposed to war but less in favor of marijuana; they are more in favor of equality for gays and lesbians but less in favor of sexual liberation generally. Marriage, motherhood, divorce, labor participation, socioeconomic status, and other factors contribute to women's voting patterns.

Many studies have probed the differences more deeply. Social psychologist Alice Eagly has studied this extensively; she pointed out in a 2009 overview of her life's work that women and men are both helpful to others, although in somewhat different ways. But she and her colleagues had already done a large combined analysis of many studies and come to certain conclusions. First, women consistently outscore men on politically relevant social compassion; this means they "support the provision of social services . . . including housing, child care, educational opportunity, and financial support in the form of welfare. Women are also more opposed to violence, including warfare, the death penalty, and partner violence, and advocate protections from violence, such as gun control." They are also more favorable toward equal rights for gays, lesbians, and, predictably, women. So the gender gap in voting is consistent with the psychological research.

However, "Women also advocate more restriction of many behaviors that are traditionally considered immoral," including casual sex and pornography. None of this changed much over the course of the late twentieth century. The gender gap in social compassion was evident in many studies over the period from the early 1970s to the late 1990s, during which both women and men fluctuated in this measure, but without shrinking or enlarging the gap. The same can be said about traditional morality, but in the other direction. Probing more deeply, a commitment to equality as a core value predicted both social compassion *and* traditional morality, *independent of sex.* This does not translate simply into a left-wing agenda, but it suggests that as women continue to participate in politics in the advanced countries and increase their role in the developing world, there will be more equality and social compassion and less tolerance for pornography and casual sex.

There are other ways to see big trends. In 2006 political scientists Torben Iversen and Frances Rosenbluth analyzed the gender voting gap in ten countries—Australia, Britain, Canada, France, Germany, Ireland, Norway, New Zealand, Sweden, and the United States—in relation to work, marriage, and motherhood. They had two measurements of attitudes. The first asked how much the respondent favors or opposes the government (1) financing projects to create more jobs, (2) reducing the workweek with the same goal, and (3) actually providing jobs for all who want work. The second simply asked whether the person supports or affiliates with left or center-left parties. It turns out that unmarried women with full-time jobs have a bigger positive gender gap on these measures than married women with full-time jobs, but the only women who have a reverse gender gap are married women who are not in the labor force. Iversen and Rosenbluth also detect time trends and project them forward: "Given the overall trend toward more women in the workforce, we are not surprised to find that women as a group seem to be moving to the left politically." Time-series analysis clearly shows such a move in rich democracies. However, because women started to the right of men, and because they are not yet fully integrated into the labor force in many countries, the complete transition may take some years.

As we saw from the Eagly analysis, it seems likely that future increases in women's influence on women (among other things) will expand government programs that provide jobs and increase equality.

If you think that's a threat to freedom, don't worry, because women's empowerment also predicts democracy. Population scientist Paula Wyndow and her colleagues did a stunning paper on this subject in December 2013. They began:

> In the latter part of the 20th century many countries moved away from autocratic rule toward more democratic regimes. During this period women's economic and social rights also improved, with greater access to education and employment, and a worldwide fall in fertility rates. The general presumption has been that democracy leads to improvements in these aspects of gender equality. However, insufficient attention has been paid to the possibility that a causal relationship may operate in the opposite direction.

Insufficient no longer; Wyndow's group analyzed data for ninety-seven countries followed from 1980 to 2005, so they could look at all the relevant dimensions and how they changed over time. While you can't do a randomized experiment on the whole world, this is a pretty good way to get some idea of what causes what.

First (in line with a lot of other data), there was a big shift, with the number of democracies more than doubling from thirty-seven to eighty in that quarter century. The countries that transitioned to democracy had higher average levels of women's educational attainment, more women in the labor force, and lower fertility—*first.* These three factors predicted more democracy five years later, and even more democracy ten years later. "Rather than being a natural consequence of economic development, we have shown empirically that female empowerment has a causal effect on democratic development, independent of the commonly used measures of modernization, and as such it deserves much greater attention in future democracy research." This is a brand-new discovery, and it solves, for these two huge changes, the problem of the chicken and the egg. Women's empowerment comes first, and democracy follows.

So if you're wondering what the future holds, these studies give us a crystal ball, and here is the shape that emerges as we gaze into the glassy globe's statistical clouds: as women gain in influence, all else being equal, the world will become more democratic, more socially compassionate, more equal, less discriminatory, less sexually casual, and less pornographic.

Those are the trends we can forecast from the grass roots up, but that's not all. Growing evidence shows that women leaders operate differently from men. Consider a recent example. The government shutdown of October 2013 ended despite a complete congressional impasse. On October 14, 2013, the *New York Times* explained how:

> In a Senate still dominated by men, women on both sides of the partisan divide proved to be the driving forces that shaped a negotiated settlement. The three Republican women put aside threats from the right to advance the interests of their shutdown-weary states and asserted their own political independence.

Senator Lisa Murkowski, Republican of Alaska, said, "I probably will have retribution in my state. That's fine. That doesn't bother me at all. If there is backlash, hey, that's what goes on in D.C., but in the meantime there is a government that is shut down. There are people who are really hurting." Two women Democrats in the Senate followed their Republican colleagues' lead, and men on both sides also followed:

> Of the 13 senators on a bipartisan committee who worked on the deal framework, about half were women, even though women make up only 20 percent of the Senate. Senator John McCain of Arizona joked at several points in their meetings, "The women are taking over."

McCain had served in the Senate since 1987, when Nancy Kassebaum was the only incumbent woman senator (Barbara Mikulski entered with McCain.) One of the male Democrats, Senator Joe Manchin III, thought that the even gender mix explained the committee's success. But Republican Senator Susan Collins, who had started the process by courageously calling for the compromise on the Senate floor, said, "I don't think it's a coincidence that women were so heavily involved in trying to end this stalemate. Although we span the ideological spectrum, we are used to working together in a collaborative way." While many of their male colleagues crossed their arms and sulked, women from opposite sides of the aisle phoned and e-mailed each other nonstop. The men saw a deal they could live with and followed suit.

This would not have surprised Senator Kirsten Gillibrand, Democrat of New York, who had said in an interview two years earlier, "My own experience in Congress is when women are on committees and at hearings, the nature of the discussion is different, and the outcomes are better—we reach better solutions, better decisions are made." So the Senate women are punching far above their weight in solving the most critical problem facing the nation, the one that must be solved before any others can be: D.C. deadlock. Incidentally, of the five women who led the historic compromise, three have children, as do Senators Amy Klobuchar, Heidi

Heitkamp, and Jeanne Shaheen, who joined their colleagues in forging the deal on the bipartisan, bi-gender committee.

Of the record twenty women sworn in as U.S. senators in January 2013, nineteen were interviewed by Dianne Sawyer on ABC News. Barbara Mikulski set the tone by criticizing the gridlock fostered by their male colleagues, saying, "We don't believe in the culture of delay." Dianne Feinstein said, "You know, we're less on testosterone. We don't have that need to always be confrontational." Gillibrand said, "If this Congress were 50 percent women, you can bet your bottom dollar we would not be debating birth control." Kelly Ayotte told a story about her eight-year-old daughter asking her not to run for president. Why? "Mom, because I want to be the first woman president." To which Claire McCaskill replied, "Did you break the news to her we're not waiting that long?"

What happens when women are in an executive rather than a legislative position? There aren't (yet) enough women heads of state to compare them systematically to men, but there are enough in some other governing roles. In what has to be the most elegant study of its kind, political scientist Lynne Weikart and her colleagues surveyed 120 mayors—65 women and 55 men—of comparable cities with populations over thirty thousand. The two samples were similar in many ways, but there were key gender differences. "Female mayors were far more willing to change the budget process, be more inclusive, and seek broader participation. Finally, more women mayors than men were willing to admit fiscal problems and discuss changes in their goals."

These findings confirmed previous studies of women in various governmental roles. The late governor Ann Richards of Texas is quoted as saying, "The joy of having power is being in a position of distributing it, giving it away, empowering others." It's hard to imagine a man describing governance as the joy of giving power away. Weikart called her article "The Democratic Sex." Interestingly, women mayors wrote much more than men in response to open-ended questions;

this is consistent with the finding in other studies that women govern more transparently. In answer to one of those questions, a woman mayor wrote, "Frequently mothers will introduce me to their daughters," a comment that speaks volumes about the future.

What about leadership in the private sector? In December 2013 it was announced that the next CEO of General Motors would be Mary Barra, a thirty-three-year veteran of the company who started at age eighteen, enrolling in GM's school for engineers. Her father had worked there as a die maker for thirty-nine years, and she herself interned on the factory floor. She was known as a "car gal," and she was the board's unanimous choice. According to the front page of the *New York Times,* "She has been a rank-and-file engineer, a plant manager, the head of corporate human resources and, since 2011, the senior executive overseeing all of G.M.'s global product development. And she has, in the parlance of the Motor City, gasoline running through her veins." Her mentor and predecessor in the job said, "Mary was picked for her talent, not her gender," and described her selection as an emotional moment: "It was almost like watching your daughter graduate from college."

The *Times* went on to say, "She is known inside G.M. as a consensus builder who calls her staff together on a moment's notice to brainstorm on pressing issues. . . . She has a soft-spoken manner that belies her intensity on the job." She is married and has two children, fourteen and sixteen. But she loves the boy stuff and has often been seen at the test track, driving at high speeds to get a feel for how new models perform. And she likes to take the boys out of *their* comfort zone; she had engineers working in car dealerships to get a feel for what customers want. As for the competition, she intended and intends to win.

She is no token example. Women now occupy many senior positions at GM, and they understand their female customers. Four who were on an engineering team developing crossover utility vehicles argued successfully for a space below the left footrest to accommodate drivers in high-heeled shoes. "Things have changed dramatically," Barra said in an interview. "There are women in every aspect of the business. You start with our board. We were just recognized for having a very high percentage of women board members. . . . Through my whole career, I have seen more and more women. It's been a decade where we have had many women plant managers. When I started in 1980 that . . . wasn't necessarily even comprehended. And maybe 15 or 20 years later there were many women across the country and across the globe in leadership roles, running operations."

It was a proud moment for her and for women, but she was not destined for an easy path. In early 2014, Barra had to begin shepherding GM through its greatest challenge in years, the revelation of a faulty ignition system that had caused at least thirteen deaths and required a massive recall of vehicles. Although there were many serious questions, she was not considered responsible for the failure, which included long delays in making the information public, and she was generally praised for her openness and poise in handling severe criticism after the fact. She took full responsibility on behalf of the company and immediately set about making changes, but to say that she was leaned on by journalists during this period would be putting it mildly.

In June 2014, the results of "a sweeping internal investigation" led by a former U.S. attorney were announced in a large-font lead article on the front page of the *New York Times.* They included widespread incompetence and neglect in the engineering and legal ranks, although no criminal actions were found and no blame was placed on Barra or her closest associates. She fired fifteen employees, including a vice president and a top lawyer, and disciplined five more, "highly unusual," noted the *Times,* "in an industry where such purges have been rare." Summarizing the report and the actions following at a meeting of around a thousand employees in Michigan, Barra said, "I never want to put this behind us. . . . I want to keep this painful experience in our collective memories."

The company had by then recalled 2.6 million vehicles and engaged the independent expert Kenneth Feinberg to develop a compensation plan for victims and their families. However, Senators Claire McCaskill and Richard Blumenthal said they would hold further hearings, and it was likely that prosecutors will continue investigating the deaths. Many lawsuits were mounted or threatened. But Barra's response to this crisis, occuring immediately after her appointment to the top job and unprecedented in the company's history, was generally thought highly professional, open, decisive, and wise. By September she had made the changes and was looking toward the future, announcing that the first GM car to drive itself would be available in two years.

You could argue, of course, that a male CEO might have done the same; so what difference will it make when there are many more women in this kind of role? There are actually a lot more systematic studies of gender and leadership in the corporate sector than there are in government.

First, surveys simply try to get at people's judgment: Who is a good leader? In 2012 business leadership consultants Jack Zenger and Joe Folkman surveyed 7,280 managers and executives, "a sampling of male and female leaders from high-performing companies" known for commitment to leadership development. The sample was predominantly male, yet the researchers rated women as better leaders in twelve out of fifteen domains. Women were seen as better in sales, marketing, human resources and training, general management, finance and accounting, product development, legal, engineering, information technology, research and development, and (slightly) quality management. Men were considered better leaders in customer service, facilities management and maintenance, and administrative and clerical domains. In the index of overall leadership effectiveness, an "average rating from an aggregate of manager, peer, direct report and other ratings," based on forty-nine leadership items, women were voted superior to men at a very high level of statistical significance.

As for specific competencies, women outranked men on twelve out of sixteen. This included the ones identified by conventional wisdom—developing others and relationship building, for example. But the biggest gaps favoring women were in taking initiative, practicing self-development, integrity and honesty, and driving for results. Men were rated slightly but significantly more positively on only one measure, "develops strategic perspective." This research was reported on in many media outlets, often with the simple phrase "Women do it better." In one interview Zenger said, "It is a well-known fact that women are underrepresented at senior levels of management. Yet the data suggests that by adding more women the overall effectiveness of the leadership team would go up."

So much for the overall ranking, made by a majority male sample. But there are more specific studies, not only of executives but also of directors. An ingenious one published in 2013 by economists David Matsa and Amalia Miller asked the key question in the title: "A Female Style of Corporate Leadership?" They took advantage of Norway's initiative a few years earlier: the country had passed a law requiring boards of directors of public companies to include women. In fact, Norway mandated a quota of 40 percent women on every board—a doubling of the proportion in the average firm—within two years. Matsa and Miller compared these companies with otherwise similar Nordic firms.

There was no significant effect on most corporate decisions. "Sizable differences emerged, however, in these firms' employment policies. Specifically, firms affected by the quota undertook fewer employee layoffs, causing an increase in relative labor costs," with a consequent reduction in profits of 4 percent compared to other firms. However, the researchers estimated the impact on employment to be much larger, around 30 percent. The authors concede that their follow-up was not yet long enough (the mandate was passed in 2006) to establish whether skipping mass layoffs might eventually drive profits for the companies, which showed loyalty to trained and tested employees. "Ultimately, time will tell whether the gender quota created value for these firms in the long run." The parliament of the European Union voted overwhelmingly in November 2013 to impose the same quota for corporate boards of large companies throughout the E.U. by 2020, so there will soon be many more opportunities to analyze the consequences.

Meanwhile, the same research team was planning the 2014 publication of an American study with related implications. They examined data for over 2,000 women-owned and nearly 48,000 other U.S. firms (all privately held) between 2005 and 2009, a time when large numbers of workers were laid off because of the Great Recession. They found that "workforce reductions are more than twice as frequent at male-owned firms as at female-owned firms (13.5% versus 5.9%)" and that average year-over-year workforce reductions also differed "substantially" (2.3 versus 1.8 percent). (Economists, probably seeking "neutral" language, call this "labor hoarding.") The gap persisted after controlling for many differences between the two groups aside from gender. As in the Norwegian study, it is not known how these differences will affect profits in the long run. Even the underlying motives remain somewhat speculative, but the authors quote Nicola Leibinger-Kammüller, the CEO of TRUMPF (a high-tech international manufacturer of machine tools and medical technology), who said on the *PBS NewsHour* in 2012,

> It's just a terrible thought having to lay off people, because we like our employees and we need them. And they are well-trained, and they're loyal. And they have been working for us for decades, some of them, or many of them have. And it's just a terrible thought to have to send them away.

They also note what Susan Spencer, a meat-processing entrepreneur and former NFL general manager, told Reuters in 2011: "Women's empathy enables them to look at business issues through a wide angle lens."

Another pair of economists, Renée Adams and Patricia Funk, surveyed both board members and CEOs in public companies in Sweden, to probe possible differences in values between men and women in these roles. They received responses from 628 people, around a third of those approached. The results were striking, even at the top:

> Male directors care more about achievement and power than female directors, and less about universalism and benevolence. This is consistent with prior literature that has found that across cultures men consistently attribute more importance to self-enhancement values (achievement and power), whereas women emphasize self-transcendence values (universalism and benevolence).

These differences in leaders parallel those often found in studies of ordinary women and men. But not all the sex differences followed the usual male-female pattern: "Female directors are less security and tradition oriented and care more about stimulation than male directors." Surprisingly, but in line with the economists' finding about security, women directors are also slightly more "risk loving" than their male colleagues. So adding more women to boards in countries that *don't* have quotas might be expected to decrease achievement orientation, power brokering, and commitment to traditions among directors, while increasing benevolence, universalism, and risk taking. By the way, women board members in Sweden are more likely to be married and have on average more children than their U.S. counterparts, probably because the social supports for mothers in Sweden—one of the world's most gender-equal countries—are much better, and the supports include paternity leave for fathers.

Interestingly, women directors across countries tend to be younger than their male counterparts, in all likelihood because it's (so far) been harder to find qualified women. The age difference is consistent with my prediction that gender balance will be easier to achieve as the pool of appropriately experienced women increases. That pool grows fast as you go down the executive ranks. But this no longer means that women are being kept down. The larger cohorts of women who have the necessary experience are moving relentlessly up. The glass ceiling is becoming a golden ladder.

What about the dark side of corporate behavior? In December 2013, sociologist Darrell Steffensmeier studied 83 real corporate frauds involving 436 defendants. A very small number of the defendants were women, some involved only through an intimate relationship with a male defendant. A few women committed fraud independently, enabled by their roles in company finance. No women were involved in conspiracies (such as that of the infamous Enron executives); women had minor roles and profited less from their crimes. The authors conclude, "Sex segregation in corporate criminality is pervasive. . . . Our findings do not comport with images of highly placed or powerful white-collar female criminals." They consider the obvious hypothesis that men have more opportunity to commit these crimes but argue that since women now make up about half of mid-level management, they should theoretically make up half the criminals at that level, instead of only 25 percent. At the upper levels, women have 15 percent of the jobs but commit only 8 percent of the crimes. In most cases they were acting with men and not independently or with other women. So women are greatly underrepresented in crime even after taking account of their more limited opportunities.

This research is consistent with the finding published by economists David Dollar and his colleagues in 2001, showing that cross-nationally, the more women there are in a country's parliament, the less corrupt the country is. They cite "a substantial literature in the social sciences which suggests that women may have higher standards of ethical behavior and be more concerned with the common good. Consistent with this micro-level evidence, we find that at the country level, higher rates of female participation in government are associated with lower levels of corruption."

They of course accept that women should play a greater role in government for the sake of fairness, but they add that "our results suggest that there may be extremely important spinoffs stemming from increasing female representation: if women are less likely than men to behave opportunistically, then bringing more women into government may have significant benefits for society in general." Corruption being one of the most intractable obstacles to progress in the developing world—the great majority of humanity—the future of the species may depend in part on the number of women in parliaments. Given that corruption has been one of the great obstacles to economic development in Africa (among other places), it has been deeply encouraging to see a dramatic increase in the percentage of women in parliaments on that continent during the past few years—often because of quotas, at least initially. Rwanda leads the world with 64 percent women representatives, and a dozen other African countries have 30 percent or more. Gretchen Bauer has summarized evidence that this trend is intensifying and spreading and tends to enhance both the position of women and the quality of democracy generally in sub-Saharan Africa.

These kinds of studies, not just in Africa but throughout the world, leave little doubt about what is happening to women and men, and not much more doubt about what the consequences will be. At the grass-roots level, the education of girls and the empowerment of women promotes, within ten years, the democratization of countries, as well as smaller families, better health for women, children, and husbands, more education for both boys and girls (with a multiplier effect for future generations), a less burdened natural environment, and more prosperous families and nations. The gender gap in voting means that as women increasingly come to the polls, governments will tend to reduce inequality, improve health and education, reduce family size, combat pornography and abusive sex, and diminish male supremacy.

More women at the top will mean governance that is more communicative and collaborative in legislatures, more candid, transparent, and open to change in executive offices, and much less corrupt. In the private sector, more women on boards of directors will make corporations less friendly to personal achievement and power and more open to universalism and benevolence. They will be somewhat less risk-averse and less committed to company traditions, show more concern about environmental protection, and be more willing to incur small losses in short-run profits to avoid mass layoffs. And when there are more women in top executive roles (CEOs, vice presidents, middle management, and so on), leadership will be more effective overall, with improvements in about 80 percent of corporate divisions and leadership skills, but especially in taking initiative, practicing self-development (this is about self-improvement, not power), integrity, honesty, and driving for results. Corruption and, especially, conspiracies to commit corruption will dramatically decline. Men will face difficulties as they adjust to this new world, as boys have long had to do in the female-forward world of school. But with a certain amount of care they will adapt and will reap many benefits.

First, the mass of men at the bottom of the pyramid will gain from it being less steep, so that the weight above them will be easier to bear. Second, they will live in a less violent world and, therefore, the weaker majority of males will not become victims of the blood sport that has dominated history. Third, boys and men will be allowed to be themselves and not constantly pressured to live up to a typecast masculinity; those with a strong nurturing side who want to share the care of children or even largely take it over, and to live more of their lives in private rather than in public will not be ridiculed. Fourth, just as the rise of women puts to rest the bad old stereotypes of female sexuality—not just that it is less driven (true on average), but that it is sweet, adorable, passive, virginal, and meekly admiring of men—so it will soften the stereotypes of male sexuality, lessening the pressure on men to perform or to prove themselves by "scoring," sequestering, dominating, or otherwise controlling women. Gays, lesbians, bisexuals, transvestites, and transgender people will stop having to hide who they are.

Now, many objections can be raised to these predictions. Women in the studies done so far have felt they had a lot to prove, have been minorities in their organizations, and have had lots of men around to mentor, help, balance, and complement them. These studies can't say for sure what women will be like when they are done with the struggle for acceptance and when their power and responsibility truly equal those of men. They certainly don't tell us what would happen in an unlikely future world *without* men. And of course, some women who come to power will be in the mold of former Alaska governor and vice-presidential candidate Sarah Palin—gun-toting women who try to out-macho men and care little for women's rights. But they are a minority, and women voters in general don't seem to like them. In any case, we can have only so much certainty about the future. If you are going to make any predictions at all, one of your best bets is going to be on the one in my admittedly flawed crystal ball: empowering women will make a better world.

# Epilogue: #YesAllWomen

In late May 2014, a deeply disturbed twenty-two-year-old man murdered four young men and two young women on his college campus in Isla Vista, California, before killing himself. He was typical in one way: of eighty mass killings involving guns between 1984 and 2014, men perpetrated seventy-eight. But he was unusual in being completely explicit about his motive: his resentment and hatred of one class of people. He left voluminous videotapes and writings in which this motive was unambiguously stated. He was launching a "war on women." In one representative video segment he said,

> Girls have never been attracted to me. I don't know why you girls aren't attracted to me, but I will punish you all for it. It's an injustice, a crime, because, I don't know what you don't see in me. I'm the perfect guy. I'll take great pleasure in slaughtering all of you. You will finally see that I am, in truth, the superior one, the true alpha male.

The "perfect guy" who will take great pleasure in slaughtering women is the "true alpha male"—a chilling use of the language of biology and a thoroughly twisted, yet not entirely false, view of human evolution. Also reflecting history, his frustration focused on women, but he killed even more men.

Rebecca Solnit, an author and historian, in her 2014 book *Men*

*Explain Things to Me,* puts it this way: "We have an abundance of rape and violence against women in this country and on this Earth, though it's almost never treated as a civil rights or human rights issue, or a crisis, or even a pattern. Violence doesn't have a race, a class, a religion, or a nationality, but it does have a gender." The perpetrators are male.

There are varying accounts of how the hashtag #YesAllWomen came about, but according to the *New York Times,* it was a response to #NotAllMen, the gist of which was that not all men are misogynistic murderers; the response was no, of course not, but *yes, all women* are and need to be afraid, really afraid, of that minority of men. There were a million #YesAllWomen tweets within a few days. Here are some of those early ones:

- #YesAllWomen because "I have a boyfriend" is more effective than "I'm not interested"—men respect other men more than my right to say no
- Because I've already rehearsed "Take whatever you want, just don't hurt me."
- because every time I try to say that I want gender equality I have to explain that I don't hate men.

And here are some that appeared or were retweeted in the first week of June:

- Because what men fear most about going to prison is what women fear most about walking down the sidewalk.
- Here's to the day when I can pump gas at 10pm and not be absolutely terrified.
- Because a smile back can get you followed, but no response can get you killed. And neither one guarantees the harassment ends.

It's impossible to read even a sampling of these anguished messages without understanding how the weight of a thoroughly male-dominated human past continues to oppress and intimidate women, how the peculiarly male viewpoint on relations between the sexes has distorted and still distorts every woman's life every day. But what I would like to do now is turn the hashtag around, to make it positive: "#YesAllWomen are equals, not just objects of men's desire."

Or maybe, when all is said and done, better than equals. How about a world in which women mainly rule? Or even, on some more distant day, a world without men?

For the first time in the history of life on our planet, evolution may soon begin to be directed by something other than natural selection; reproduction is coming under human control. Dame Sally Davies, the chief medical officer for England, announced in June 2013 that nuclear transfer had been approved for the United Kingdom and, pending parliamentary debate, might be implemented before the end of 2014. The procedure, which would be offered to couples at risk for a mitochondrial genetic disease, involves taking an egg from a donor, removing its nucleus, replacing that with a fertilized nucleus from the couple, and implanting the new egg in the woman from the couple at risk. It can also be done by transferring an unfertilized nucleus from the woman to the donor's egg, then fertilizing it with the man's sperm and implanting it.

Mitochondria are our energy centers, and they have their own genes. But sperm don't have them, so they are passed down only through women. If a woman has a mutation in her mitochondrial genes, she will pass it on to her children. But if (as in this procedure) her and her husband's other genes are transferred to another woman's egg, the donor's normal mitochondria will be passed on instead. The child will in effect have three genetic parents, a mother and a father, sharing the vast majority of the child's genes, but also a second, mitochondrial mother.

This is now. Next, consider a modified version of nuclear transfer in which the nucleus is removed from the egg of one woman (say, in a lesbian couple) and used to fertilize the egg of another. This child would have two mommies, but not just in the sense of the children being brought up by lesbian couples today; she—and the child would always be a she—would have two *biological* mommies.

The end of men?

Well, we can also imagine that the sperm of the two men in a gay couple, one bearing an X chromosome and the other carrying either an X or a Y, have their DNA extracted and used to replace the DNA in a donor egg. The thing is, they need the donor egg. Oh, and by the way: they need a womb.

Surrogacy could become more common and meet this need, but this path can't lead to the end of women. Eventually, I suppose, we can imagine a completely external human gestation. We are able to keep babies alive from earlier and earlier fetal stages—we're down to about half of pregnancy—but that's with a high risk of death and disability. So external gestation is a very long way off, but I suspect that the union of nuclear DNA from two women is at most decades away, and there will be other ways of doing it besides nuclear transfer—for example, by reverse-engineering women's skin cells to form a germ cell and then an egg and a sperm. So if one sex is going to decide to eliminate the other, women will have a jump on men by at least a couple of centuries. Besides, the decline of males may already be under way.

There are biological and social clues. The Y chromosome has shrunk in size over the eons, although not much in the last 25 million years, and a 2014 study suggests it can't shrink much further. But human sperm counts have gone down in a time span of decades, and this is likely ongoing. Lionel Tiger's *The Decline of Males* pointed to this trend in 1999. Research in France has shown a decline of about a third between 1989 and 2005. A Finland study showed a substantial decline, and a Denmark study showed an *increase* in sperm concentration (potentially good for making babies, since the egg is more densely surrounded) but a decrease in *quality* and likelihood of fertility. A study in India also showed increased concentration but decreased motility. The consensus seems to be that some sort of decline is real, and various proposed causes include hot baths, sedentary lives, tight underpants, fatty foods, maternal smoking, and global warming.

What about other traditional contributions of men, aside from DNA? Men's physical strength is needed less and less, as manufacturing wanes, robots take over more and more factory tasks, and service economies grow. In less developed countries that have not made this transition, women do hard physical labor, too. In war, the use of drones, bots, and other technologically advanced weapons systems will continue to limit the need for physical strength. In very specific situations—hand-to-hand infantry combat, special forces operations like the one against Osama bin Laden, and carrying people out of burning buildings or off battlefields—muscle matters, but in many situations it no longer does.

Children are certainly learning to see males as dispensable. About 40 percent in the United States are born to unmarried mothers, and many more will spend some time in single-parent households, four out of five headed by women. The percentage of married women who outearn their husbands quadrupled from 6 in 1960 to 24 in 2011, and the graph is a straight line going upward; the statistics are similar for married couples with children under eighteen. As for education, the percentage of couples with children where the wife is more educated than the husband went from 7 to 23 in the same time period, while the percentage where the reverse was true hovered around 16 (in most couples, husband and wife are about equally educated). Given the differing rates of high school and college graduation and the impact of those degrees on earnings, the educational disparity favoring wives over husbands can only increase.

Contrary to some claims, however, there is little evidence that to grow up mentally healthy a child must have a mommy and a daddy in an uninterrupted stream for eighteen years. Children who have two mothers or two fathers differ only in that they grow up on average less homophobic than conventionally raised kids (although some data suggest that gay fathers are more involved parents). Divorce makes life difficult for some children but saves others, and if they don't experience (1) a big drop in standard of living, (2) a complete loss of contact with one parent, or (3) continuing open conflict between the parents, they do as well as kids with non-divorced parents, some of whom probably shouldn't have stayed together. Children of single parents are more disadvantaged, and it's important for them to have at least one person besides the parent they can relate to and trust, but if they have that, along with a certain amount of resilience, they will do well. So it is likely that in a world of women, where women maintain strong social networks and relationships, children could grow up very nicely, thank you—especially if they are only girls.

But the end of men? Really?

What we are actually talking about is directing evolution, and that will become easier and easier. Evolution, finally, is about changes in genes, and we are steadily taking charge of them.

We saw in chapter 2 that whiptail lizards evolved all-female species from ancestors that had both sexes. Reproduction in these whiptails preserved the sexual interactions—and most likely the sexual pleasure—of their two-sex condition yet dispensed with males, along with all their baggage. Females take turns having offspring and serving the purpose of stimulating ovulation and triggering development as males once did with their bodies and their sperm. But women might not want to wait a million years. They could instead deploy biological science on behalf of male-free reproduction—slowly enough, perhaps, to accustom themselves to taking pleasure in each other as the males dwindle, partly naturally, partly by design.

Or perhaps they could follow the jacana route, engineering males to be smaller and, with the help of sperm selection, more numerous. This might prevent the downside of jacana females' lives: fighting viciously and killing one another's young in order to commandeer scarce males. Men might have to be modified a bit to make them better full-time fathers and prevent them from becoming violent, as surplus males tend to do. Something would have to be done about men's sex drive; although perhaps women could learn to be happy in a newly polyandrous world, where they need to have sex with numerous men in rotation, to keep the boy-toy harem more or less calm. For this to work, you wouldn't want to make the men *too* small—although chemical adjustment of women's sexuality, by women, would by then be quite advanced. Eventually, guided evolution could include external gestation after the first few months of pregnancy. That way, women could reproduce fast enough to keep the pint-sized fathers happy and busy.

Or we could aim for the lemur syndrome: women the same size as men or a little bigger but dominating decisively, the weakest woman feared by the strongest man. Polyamory would be the order of the day, with males still competing for access and females the deciders, but both sexes would get plenty of variety and males would ultimately stay diffident, hopeful, and respectful. This pattern would recover an early era of primate evolution, before monkeys and apes emerged and took new paths, some leading to extreme male domination.

Recall, however, that not all monkeys went that way; for another model we could also turn to our cousins in South America, the marmosets, with their imperfect but stable pair bonds and utterly devoted dads, carrying their twins around all day except when the babies sidle over to Mom to suckle. Sometimes there are two males to a female, sometimes an extra female—it varies; it's a mellow, share-and-share-alike sort of world. Another century's humans could learn from them.

In a more distant future, evolution could go much further, as some species have done, although ours would be under our own—or at least women's—control. Men reduced to diminutive parasites that sink their teeth into women's sides and fuse with them, delivering a periodic pulse of sperm? Somehow I don't see women choosing that, although it does fulfill the age-old romantic goal of two becoming one. As for the black widow or praying mantis solutions, women would have to change their appetites pretty drastically—but culinary fashion does evolve.

Most simple would be the gradual solution. Beginning a few decades from now, with more and more women choosing artificial insemination and single motherhood (or partnerships with other women, sexual or not), they may increasingly choose to have their own eggs fertilized by the DNA from another woman's egg. This would very slowly increase the numbers of women relative to men, which would be a lot less risky than the excess of men we see in some countries today. In this female-forward world, women would control the hookup culture.

The very slowly shrinking numbers of men should be quite happy being outnumbered, as long as some women still want to have sex with them—and as long as they behave themselves. But over many generations, certainly over evolutionary time scales, we could theoretically see men fully replaced or literally kept in small numbers for sexual services, as well as for heavy lifting and changing lightbulbs on high ceilings. The men, in this future world, could easily be engineered to have broad shoulders, square jaws, sexual prowess, amorous sensitivity, made-to-order intelligence, a sense of humor, or whatever other features women decide on. Women could create the sperm-bank book in advance, then either use the sperm or just use the men. Certainly, they could scientifically limit the men to minimal levels of aggression, arrogance, self-importance, and entitlement. With one man sexually serving several or many women, he will be enjoying life, and they won't have to have sex with him except when they ring the bell. Or a group of discerning women could choose to share several men: a beefy blond, a tall, dark, handsome brunet, a weight lifter, a marathoner, an androgynous one for babysitting, a long-haired hunk for wild sex on demand—you get the idea.

I freely admit that these are all bio-fantasies, but some of them, at least, will be possible in the future. Will women want them? A friend who read a draft of this book commented, "As a woman I hate the idea of a manless world, and I think that 98 percent of women would agree." I hope she's right, since the end of men would mean the end of billions like me. Men, many women say, do add variety and even excitement to their lives (when we're not behaving too badly), and *vive la différence* seems to point to a long two-sex future.

Maybe it's just my male desire to stick around and be a part of it, but I think the most logical route to that future would be to bonobo-ize humanity. We have two closest relatives, it's true, but this could be our chance to toss out the "battle" part of relations between the sexes—the part we share with chimpanzees: male domination, physical aggression, rape, and murderous men-in-groups-type raids. Instead we could guide our future evolution in the way of our equally close bonobo cousins. This could lead to unshakable female coalitions, based perhaps in part on sex, and males who are not unhappy (males in that female-forward species actually have a great life) but never get out of hand. They don't just know their place; they make love, not war. It's not that they're incapable of aggression; it's that they keep a lid on it and never use it on females. As in the lemur syndrome, polyamory for all could be a dividend, including the pleasure females take together, although monogamy would remain an option, and current studies suggest that women will vote for it for the foreseeable future. But, heterosexual or not, bonobos' unique face-to-face sexual intercourse can produce what look like big smiles.

In the end, the simplest solution is staring us in the face, and it doesn't require genetic tricks: go back to the rules that prevailed among hunter-gatherers for 90 percent of human history. That means women and men working at their jobs, sharing, talking, listening, and taking care of children, their main link to the future. Men didn't strongly dominate because they couldn't, whatever their motives; women had a voice because they were always there, and their contributions were critical—not just in child care and in bringing home the vegetables that were the reliable staples but in speaking truth to would-be male power every night around the fire. Under certain circumstances, both polyandry and polygyny were available options. There was violence and it was mainly male, but it was mostly random and interpersonal, more like an accident than an ideology. We needn't imitate hunter-gatherers slavishly; men should be able to do a lot more child care than typical hunters do. But at a minimum hunter-gatherers serve as a *human* model, showing us what is in our psychological scope without genetic change, and they probably also tell us something about what we were like for most of human history.

Current technology, as we've seen, has mostly replaced men's muscular strength and is making inroads even into martial prowess; more importantly, it is creating a level playing field again in communications. Women's coalitions now are vast webs in cyberspace—the majority of active Facebook users are women. As COO Sheryl Sandberg has said, "The world's gone social, and women are more social than men." The campfire is on the screen and in the social networks. Once again we are blurring the lines between our homes and our public spaces, between life, business, and politics, between a falsely diminished female and a falsely aggrandized male. Once again women's voices really matter. With a clever combination of technology and open-minded experiment, and without pretending that men and women are the same, we can use this chance to try real equality.

According to a bad old adage in Lesotho, "A woman is the child of her father, her husband, and her son." This sequence was long true in Western culture, too, and was highlighted in 1869 by Elizabeth Cady Stanton as "invading . . . our homes, desecrating our family altars, dividing those whom God has joined together, exalting the son above the mother who bore him, and subjugating, everywhere, moral power to brute force."

This is over. Women have come so far that, however far they have to go, there is no turning back. Fifty years from Stanton's speech to the vote; fifty from that to the second wave; fifty from then till now. In another fifty, equality may not be numerically, exactly here in every realm, but it will be evident to everyone that the writing is on the wall. History is on women's side. Even in the most oppressive places in the world, they will act as a deeply subversive force, sensitized to the changing lives of other women throughout the world, undermining backwardness and oppression in their own.

Anthropologist Helen Fisher, in her book *The First Sex,* was similarly optimistic: "Like a glacier, contemporary women are slowly carving a new economic and social landscape, building a new world. . . . We are inching toward a truly collaborative society, a global culture in which the merits of both sexes are understood, valued, and employed." I agree, except that only the relentlessness will be glacial; the pace will not. It will be like the rise in life expectancy, the decline of family size, the spread of democracy—a matter of decades or, at most, generations. Having known five generations in my own limited life, I don't see this as a long time. In the light of evolution, archaeology, and history, it's no time at all.

My grandparents lived nearly the entire span from the beginning of the suffragist movement until the 1950s. We don't think of that decade as a good time for women, and in many ways it wasn't, but by then it had become routine for women to vote, work, go to college, and drive, none of which was true when my grandparents met and married. I've seen for myself all the transformations from the 1950s until now. My grandson (I don't have a granddaughter yet) will see at least as much change as I did—his father is a superb model—and when he is my age, women will have gained enormously in rights, equality, and power.

They will not make a perfect world, but it will be less flawed than the one men have made and ruled these thousands of years. They will build on men's achievements and mend some wounds in the legacy—not because they are earth mothers but because they are earth citizens, with a strongly pragmatic bent and a balance of healing and damage that is, in their very marrow, different from that of males.

My grandson, I think, will be happy in the new world. It will be better for him *because* women help run it. All indications are that as leaders they will be fair to him. But during the transition, some men will feel pain and some boys will need protecting; they may be angry, as they have often been in the past. They will need sympathy, patience, education, leadership, and wisdom to bring them along, and to ensure that they do not resent the successes of their mothers, sisters, lovers, wives, friends, and daughters. But, however tumultuous, this transition will in the end make a better world for everyone. In anthropological or even historical terms, it will not take much longer: half a century, a century at most—nothing, really, when you consider the depth of the past.

We have barely begun to see what women can do. We will not see the end of men—they, too, have a contribution to make—but we will see the end of male supremacy. It will be a great beginning for women and a new start for men, too. Women are smart, determined, steady, fair, calm, strong, optimistic, capable, democratic, cooperative, and unstoppable. They are attentive to their own flaws and grateful for the strengths and contributions of others, men included. Their future history is an open book before them; the empty pages beckon with impatience, expectation, and even a hint of seduction, and women will fill those pages with splendid untold stories.

# Acknowledgments

Sex differences being a subject that interests everyone, and on which every life supplies endless information, it will come as no surprise that what I know about it rests on the shoulders of countless people—parents, teachers, friends, relatives, lovers, children, students, colleagues, and even strangers. Add to this the fact that it has been one of my academic interests for a lifetime, and you can imagine a list of thank-yous that goes on for many pages. My gratitude extends far beyond the few people I can mention here.

First and foremost, I thank *some* of the women who not only were role models and helped me believe in myself, but who also made it clear to me that women can do anything:

–Hannah Levin Konner, Dora Venit, Jean Silbersweig, Jane Chermayeff, Beatrice Whiting, Ronnie Wenker Konner, Nancy Howell, Elaine Markson, Kathy Mote, Joanna Adams, Betty Castellani, Deborah Lipstadt, Shlomit Finkelstein, Sherry Turkle, Melissa Fay Greene, Mari Fitzduff, Zoë Heller, Patricia Greenfield, Sarah Blaffer Hrdy, Polly Wiessner, Megan Biesele, Sally Gouzoules, and Carol Worthman;

–my late wife Marjorie Shostak, friend of my youth, adventurous partner in fieldwork and in life, African traveler, author of the anthropological classic *Nisa: The Life and Words of a !Kung Woman*, illuminator of women's roles cross-culturally, passionate mother of my children, unfailing supporter of my plans and dreams, and brave fighter against cancer;

–my wife and friend of many years now, Ann Cale Kruger, a dedicated and brilliant child psychologist who has taught me so much about how the environment molds us, brought the arts to disadvantaged children, worked to save girls from sexual exploitation, shaped generations of students, teachers, and counselors, and made a home for our two families in harmony;

–the many other women I have had as colleagues and friends who taught me lessons large and small;

–the women of the !Kung hunter-gatherers among whom Marjorie and I lived for two years, who showed extraordinary bravery, skill and judgment in the challenges of an intensely difficult life, while tolerating our questions and intrusions;

–and the women who admitted an anxious medical student into their illnesses, healing processes, childbirths, motherhood, pain, suffering, joy, and, when possible, to the thrill and triumph of recovery.

I also thank the men who along the way have helped make me a better man, teaching me that loving women includes admiring them, and who helped me suppress any slightest tendency I might have had to think or feel that my own success depended on treating women as anything less than equal: Irving Konner, Herbert Perluck, Lawrence Konner, Martin Silbersweig, Irven DeVore, Gerald Henderson, Stefan Stein, John Whiting, David Silbersweig, Joe Beck, Julian Gomez, Boyd Eaton, Steve Berman, Charles McNair, Misha Pless, Leslie Rubin, and Joe Weber.

I thank the countless teachers, in addition to some mentioned above, who helped me understand the processes of evolution, development, and brain function, especially as applied to understanding sex differences: Jerome Kagan, Ernst Mayr, Nicholas Blurton Jones, Richard Lee, Patricia Draper, Jane Lancaster, Patricia Whitten, Alice Rossi, Daniel Federman, Norman Geschwind, Robert Sapolsky, James Rilling, and Jared Diamond, among many others.

I gratefully acknowledge my debt to my literary agent of more than three decades, Elaine Markson, who represented and supported my work with professionalism, loyalty, and humor, and to her associates Geri Thoma, Gary Johnson, and most recently Chaya Levin. My own assistant Kathy Mote, who has been with me and my family for twenty-seven years, provided invaluable help in this as in so many other projects.

At W. W. Norton & Company, John Glusman acquired the manuscript and gave it two careful readings at different stages, improving it greatly. He also asked others at the company to read it, and their comments were extremely helpful, as was the work of John's assistant Jonathan Baker, of the superbly careful and widely knowledgeable copy editor Bonnie Thompson, and of the publicists Rachel Salzman and Lauren Opper. My experience justified the claim on Norton's website that because it is exclusively owned by them, "Norton's employees answer to each other, not to outside shareholders or directors. With a shared commitment to the company at heart, they publish only what they love to publish." It was my good fortune to come under that umbrella.

Outside of Norton, the manuscript was read in full by Faith Levy and in part by Anna Beck, Brian Diaz, Ann Cale Kruger, Jennifer Kuzara, Michael Peletz, and Julie Seaman; all made valuable comments. Tami Blumenfield and John Townsend answered important questions. Sarah Hrdy and Sherry Turkle made helpful suggestions in addition to providing pre-publication praise. I am grateful to all these friends and colleagues for their time, effort, and expertise.

Finally, I thank my children, Susanna Konner Post, Adam Konner, Sarah Konner, stepdaughter Logan Kruger, and "extra daughter" Becky Perry, for allowing me into their lives and even more for keeping me in, really in, their lives when they began to have a choice; and my grandson Ethan, not yet three but full of developmental lessons. These young people have taught me more about how we become women and men, and more importantly how we become human, than I can ever thank them enough for.

To them, and to Irven DeVore, my lifelong teacher and friend who died on September 23, 2014, and whose insights into the nature of sex differences *and* similarities pervade this book, it is gratefully dedicated. Only I, alas, am accountable for the much that must be wrong with it. For indulging my flawed attempt to shed light on matters that Darwin aptly called "hidden in darkness," I also thank you, the reader, and hope that, despite my stumbles and my dim, flickering lamp, it has been a partly illuminating journey.

# Notes

## Introduction: "Stronger Than All Besides"

vii *Elizabeth Cady Stanton speech:* "Address to the National Woman Suffrage Convention, Washington, D.C., January 19, 1869," in *The Concise History of Woman Suffrage: Selections from History of Woman Suffrage*, ed. Paul Buhle and Mari Jo Buhle (Champaign-Urbana, IL: University of Illinois Press, 2005), 249–56.

7 *"The problem of woman":* Simone de Beauvoir, *The Second Sex* (1949; repr., New York: Vintage, 2011).

9 *water striders:* L. Rowe, G. Arnqvist, J. Krupa, and A. Sih, "Sexual Conflict and the Evolutionary Ecology of Mating Patterns: Water Striders as a Model System," *Trends in Ecology and Evolution* 9 (August 8, 1994): 289–93. Video of water striders mating: https://www.youtube.com/watch?v=URjVA2CUgSo, accessed Sept. 12, 2014.

11 *"Male bonding and patriarchy":* Camille Paglia, *Sexual Personae: Art and Decadence from Nefertiti to Emily Dickinson* (New Haven: Yale University Press, 1990), 12.

## Chapter 1: Diverge, Say the Cells

19 *Herculine Barbin:* Michel Foucault and Herculine Barbin, *Herculine Barbin: Being the Recently Discovered Memoirs of a Nineteenth-Century French Hermaphrodite* (New York: Pantheon, 1980).

22 *a third sex or gender:* For an excellent broad overview of how people who are not clearly and simply male or female have been treated,

see Gilbert Herdt, ed., *Third Sex, Third Gender: Beyond Sexual Dimorphism in Culture and History* (New York: Zone Books, 1994).

25 *Here is the story of sexual development:* For a marvelous first-person account of this process, see Joan Roughgarden's *Evolution's Rainbow: Diversity, Gender, and Sexuality in Nature and People* (Berkeley: University of California Press, 2013), chapter 10.

27 *Barbin's modern counterparts:* For a readable and fairly recent overview, see Melissa Hines's *Brain Gender* (New York: Oxford University Press, 2005). For a scientific update, see her article "Gender Development and the Human Brain," *Annual Review of Neuroscience* 34 (2011): 69–88. For collections of authoritative reviews, see Jill B. Becker, Karen J. Berkley, Nori Geary, Elizabeth Hampson, James P. Herman, and Elizabeth A. Young, eds., *Sex Differences in the Brain: From Genes to Behavior* (New York: Oxford University Press, 2008), and the group of papers introduced by Simon LeVay in *Frontiers in Neuroendocrinology* 32 (2011): 110–263. Specific studies are cited below.

29 *Not everyone interprets these facts in the same way:* For a recent critique, see Rebecca M. Jordan-Young, *Brain Storm: The Flaws in the Science of Sex Differences* (Cambridge, MA: Harvard University Press, 2011).

29 *In one classic study:* C. H. Phoenix, J. A. Resko, and R. W. Goy, "Psychosexual Differentiation as a Function of Androgenic Stimulation," in *Perspectives in Reproduction and Sexual Behavior,* ed. M. Diamond (Bloomington: Indiana University Press, 1968), 33–49; R. W. Goy, "Experimental Control of Psychosexuality," *Philosophical Transactions of the Royal Society of London B* 259 (1970): 149–62.

29 *thousands of experiments:* For a review and starting point, along with some new complexities, see Kathryn M. Lenz and Margaret M. McCarthy, "Organized for Sex: Steroid Hormones and the Developing Hypothalamus." *European Journal of Neuroscience* 32, no. 12 (2010): 2096–2104, 2011; the 2008 collection edited by Jill Becker and colleagues, *Sex Differences in the Brain;* and the special 2011 issue of *Frontiers in Neuroendocrinology* cited above.

30 *A 2004 study:* William G. Reiner and John P. Gearhart, "Discordant Sexual Identity in Some Genetic Males with Cloacal Exstrophy Assigned to Female Sex at Birth," *New England Journal of Medicine* 350, no. 4 (2004): 333–41.

30 *Reiner's longer paper:* William G. Reiner, "Psychosexual Development in Genetic Males Assigned Female: The Cloacal Exstrophy Experience,"

*Child & Adolescent Psychiatric Clinics of North America* 13, no. 3 (2004): 657–74.

30 *BJ interview:* Ibid., 668.

31 *"children seemed to be almost empowered":* Ibid., 667.

31 *"demonstrates a scenario that is typical":* Ibid., 668.

31 *"These children adapt"* and *"an important role":* Ibid., 671.

32 *One such unlucky child:* Natalie Angier, "Sexual Identity Not Pliable After All, Report Says," *New York Times*, March 14, 1997, http://www.nytimes.com/1997/03/14/us/sexual-identity-not-pliable-after-all-report-says.html?module=Search&mabReward=relbias%3As, accessed Sept. 12, 2014; Milton Diamond and H. Keith Sigmundson, "Sex Reassignment at Birth: Long-Term Review and Clinical Implications," *Archives of Pediatric and Adolescent Medicine* 151, no. 3 (1997): 298–304.

32 *"He got himself a van":* Angier, "Sexual Identity."

32 *John's suicide:* Debra Black, "Sex, Lies and a Quest for Identity: The Boy Raised as a Girl Suffered for Social Experiment," *Toronto Star*, May 11, 2004, http://www.cirp.org/news/torontostar05-11-04/, accessed Sept. 12, 2014.

32 *In another such case:* Susan J. Bradley, Gillian D. Oliver, Avinoam B. Chernick, and Kenneth J. Zucker, "Experiment of Nurture: Ablatio Penis at 2 Months, Sex Reassignment at 7 Months, and a Psychosexual Follow-up in Young Adulthood," *Pediatrics* 102, no. 1 (1998): e9.

33 *growing number of transsexuals:* For a recent and authoritative multidisciplinary view, see the volume edited by Randi Ettner, Stan Mostrey, and A. Evan Eyler, *Principles of Transgender Medicine and Surgery* (Binghamton, NY: Haworth, 2007).

33 *"Many young children experiment":* Randi Ettner, *Confessions of a Gender Defender: A Psychologist's Reflections on Life Among the Transgendered* (Evanston, IL: Chicago Spectrum, 1996), 21.

33 *"Few things are as devastating":* Ibid., 28. Fortunately, this is beginning to change. A young friend of mine here in Georgia has decided to change sex and is in the process of becoming a man. Her family, which consists of fairly typical high school–educated people who grew up here, has been for the most part remarkably accepting and supportive.

34 *male-to-female transsexuals who came to autopsy:* A. Garcia-Falgueras,

L. Ligtenberg, F. P. Kruijver, and D. F. Swaab, "Galanin Neurons in the Intermediate Nucleus (InM) of the Human Hypothalamus in Relation to Sex, Age, and Gender Identity," *Jounral of Comparative Neurology* 519, no. 15 (2011): 3061–84.

34 *Dominican Republic studies:* J. Imperato-McGinley, R. E. Peterson, T. Gautier, and E. Sturla, "Androgens and the Evolution of Male-Gender Identity Among Male Pseudohermaphrodites with 5a-Reductase Deficiency," *New England Journal of Medicine* 300 (1979): 1233–70; J. Imperato-McGinley, "5alpha-Reductase-2 Deficiency and Complete Androgen Insensitivity: Lessons from Nature," *Advances in Experimental Medicine & Biology* 511 (2002): 121–31.

37 *Melissa Hines study of XY women:* Melissa Hines, S. Faisal Ahmed, and Ieuan A. Hughes, "Psychological Outcomes and Gender-Related Development in Complete Androgen Insensitivity Syndrome," *Archives of Sexual Behavior* 32, no. 2 (2003): 93–101.

38 *Studies of identical and fraternal twins:* For an excellent and highly readable recent account of the history and science of twin studies culminating in the Minnesota Twin Study, see Nancy Segal's *Born Together—Reared Apart: The Landmark Minnesota Twin Study* (Cambridge, MA: Harvard University Press, 2012).

## Chapter 2: Hidden in Darkness

46 *Komodo dragon habits:* James B. Murphy, ed., *Komodo Dragons: Biology and Conservation*, Zoo and Aquarium Biology and Conservation Series (Washington, DC: Smithsonian Books, 2002); Walter Auffenberg, *The Behavioral Ecology of the Komodo Monitor* (Gainesville: University Press of Florida, 1981). For a brief, accessible overview, see Jennifer S. Holland, "Once Upon a Dragon," *National Geographic* 255, no. 1 (2014): 96–109.

46 *Venom:* B. G. Fry and twenty-seven other authors, "A Central Role for Venom in Predation by *Varanus komodoensis* (Komodo Dragon) and the Extinct Giant *Varanus (Megalania) priscus*," *Proceedings of the National Academy of Sciences USA* 106, no. 22 (2009): 8969–74.

46 *Komodo reproduction without males:* P. C. Watts, K. R. Buley, S. Sanderson, W. Boardman, C. Ciofi, and R. Gibson, "Parthenogenesis in Komodo dragons," *Nature* 444, no. 7122 (2006): 1021–22.

47 *Asexual and sexual whiptail lizards:* D. Crews, "Evolution of Neuroendocrine Mechanisms That Regulate Sexual Behavior," *Trends in Endocrinology and Metabolism* 16, no. 8 (2005): 354–61; David Crews, Nicholas S. R. Sanderson, and Brian G. Dias, "Hormones, Brain, and Behavior in Reptiles," in *Hormones, Brain and Behavior,* 2nd ed., vol 2., ed. Donald W. Pfaff (New York: Academic Press, 2009), 771–816.

47 *Darwin's daring primroses:* Charles Darwin (1862), "On the Two Forms, or Dimorphic Condition, in the Species of *Primula*, and on Their Remarkable Sexual Relations," *Journal of the Proceedings of the Linnean Society of London (Botany)* 6 (1862): 77–96; accessed online Sept. 12, 2014, at http://darwin-online.org.uk/converted/published/1862_primula_F1717.html.

47 *"We do not even know":* Ibid., 94–95.

48 *"the twofold cost of producing males":* John Maynard Smith, *The Evolution of Sex* (Cambridge: Cambridge University Press, 1978), 58.

49 *"from the sexual, or amatorial, generation of plants":* Erasmus Darwin, *Phytologia; or, The Philosophy of Agriculture and Gardening. With the Theory of Draining Morasses and With an Improved Construction of the Drill Plough* (London: P. Byrne, 1800), 104. E-book scanned from the original, online at https://play.google.com/books/reader?id=YggpAAAAYAAJ&printsec=frontcover&output=reader&authuser=0&hl=en&pg=GBS.PP11, accessed Sept. 12, 2014.

49 *"sexual reproduction is the chef d'oeuvre":* Ibid., 103.

49 The Loves of the Plants: Erasmus Darwin, *The Poetical Works of Erasmus Darwin, M.D. F.R.S. Containing The Botanic Garden, in Two Parts; and The Temple of Nature. With Philosophical Notes and Plates. In Three Volumes. Vol. II Containing The Loves of the Plants* (London: J. Johnson, 1806). E-book scanned from the original, accessed online Sept. 12, 2014, at https://play.google.com/books/reader?id=uhs6AAAAMAAJ&printsec=frontcover&output=reader&authuser=0&hl=en&pg=GBS.PR1.

50 *"Each wanton beauty":* Ibid., 17.

50 *"It is interesting to contemplate a tangled bank":* Charles Robert Darwin, *On the Origin of Species by Means of Natural Selection, or the Preservation of Favored Races in the Struggle for Life; with a New Foreword by George Gaylord Simpson* (1859; repr., New York: Collier, 1962), 484.

51 *"Thus, from the war of nature":* Ibid., 484–85.

52 *proved true for a species of sea squirt:* J. D. Aguirre and D. J. Marshall, "Does Genetic Diversity Reduce Sibling Competition?" *Evolution* 66, no. 1 (2012): 94–102.

52 *The Red Queen:* For a good brief introduction, see Carl Zimmer's article "On the Origin of Sexual Reproduction," *Science* 324 (2009): 1254–56.

54 *The black widow:* Helma Roggenbuck, Stano Pekár, and Jutta M. Schneider, "Sexual Cannibalism in the European Garden Spider *Araneus diadematus*: The Roles of Female Hunger and Mate Size Dimorphism," *Animal Behaviour* 81 (2011): 749–55. For a general account of the evolution of sexual cannibalism, see Mark A. Elgar and Jutta M. Schneider, "Evolutionary Significance of Sexual Cannibalism," *Advances in the Study of Behavior* 34 (2004): 135–63.

55 *Orb weaver spiders prevented from eating males:* Klaas W. Welke and Jutta M. Schneider, "Sexual Cannibalism Benefits Offspring Survival," *Animal Behaviour* 83, no. 1 (2012): 201–07.

55 *"nutrient-limited"* and *"males are high-quality prey":* R. Rabaneda-Bueno, M. Á. Rodríguez-Gironés, S. Aguado-de-la-Paz, C. Fernández-Montraveta, E. De Mas, D. H. Wise, and J. Moya-Laraño, "Sexual Cannibalism: High Incidence in a Natural Population with Benefits to Females," *PLoS One* 3, no. 10 (2008): e3484, p. 1.

55 *"Safer Sex":* Lutz Fromhage and Jutta M. Schneider, "Safer Sex with Feeding Females: Sexual Conflict in a Cannibalistic Spider," *Behavioral Ecology* 16, no. 2 (2004): 377–82.

55 *Female praying mantises have their own charms:* Katherine L. Barry, Gregory I. Holwell, and Marie E. Herberstein, "Male Mating Behaviour Reduces the Risk of Sexual Cannibalism in an Australian Praying Mantid," *Journal of Ethology* 27, no. 3 (2008): 377–83. For a lovely video (not suitable for squeamish males), see https://www.youtube.com/watch?v=KYp_Xi4AtAQ, accessed Sept. 12, 2014.

56 *Octopus sexual cannibalism:* Roger T. Hanlon and John W. Forsythe, "Sexual Cannibalism by *Octopus cyanea* on a Pacific Coral Reef," *Marine and Freshwater Behaviour and Physiology* 41, no. 1 (2008): 19–28.

57 *Consider the syrupy primordial slime:* Of the many accounts of the whole of evolution, I recommend Michael Ruse and Joseph Travis, eds., *Evolution: The First Four Billion Years,* introduction by Edward

O. Wilson (Cambridge, MA: Belknap Press of Harvard University Press, 2009).

57 *a string of genetic material:* The best popular account of the logic of this—the essence of the evolutionary process—remains Richard Dawkins's *The Selfish Gene* (1989; repr., 30th anniversary ed., New York: Oxford University Press, 2006).

58 *Trich meiosis:* S. B. Malik, A. W. Pightling, L. M. Stefaniak, A. M. Schurko, and J. M. Logsdon Jr., "An Expanded Inventory of Conserved Meiotic Genes Provides Evidence for Sex in *Trichomonas vaginalis*," *PLoS One* 3, no. 8 (2008): e2879.

59 *the asexual New Zealand mud snail:* D. Paczesniak, S. Adolfsson, K. Liljeroos, K. Klappert, C. M. Lively, and J. Jokela, "Faster Clonal Turnover in High-Infection Habitats Provides Evidence for Parasite-Mediated Selection," *Journal of Evolutionary Biology* 27, no. 2 (2014): 417–28.

59 *"Ancient Asexual Scandals":* Olivia Judson and Benjamin B. Normark, "Ancient Asexual Scandals," *Trends in Ecology and Evolution* 11 (1996): 41–46.

60 *"Love is coming down to war":* Ronald Chase is quoted by Hillary Mayell in "Lovebirds and Love Darts: The Wild World of Mating," National Geographic Daily News, February 13, 2004, accessed Sept. 12, 2014, at http://news.nationalgeographic.com/news/2004/02/0212_040213_lovebirds.html. For the scientific background see Ronald Chase and Katrina C. Blanchard, "The Snail's Love-Dart Delivers Mucus to Increase Paternity," *Proceedings of the Royal Society: Biological Sciences* 273, no. 1593 (2006): 1471–75.

61 *"Sex and Violence in Hermaphrodites":* N. K. Michiels, and L. J. Newman, "Sex and Violence in Hermaphrodites," *Nature* 391 (1998): 647. The quotes from Newman and Eberhard are reported by Susan Milius, "Hermaphrodites Duel for Manhood," *Science News* 153, no. 7 (February 14, 1998): 101, accessed online Sept. 12, 2014, at http://www.readcube.com/articles/10.2307/4010187?locale=en.

62 *our old friend the unisexual whiptail lizard:* The next few pages draw on the work of David Crews and his colleagues, from "Evolution of Neuroendocrine Mechanisms."

63 *"snapshot of evolution":* Crews, "Evolution of Neuroendocrine Mechanisms," 354.

63 *"Given that the first 'sex' was female":* B. G. Dias and D. Crews, "Regulation of Pseudosexual Behavior in the Parthenogenetic Whiptail Lizard, *Cnemidophorus uniparens*," *Endocrinology* 149, no. 9 (2008): 4629–30.

64 *progesterone induces "the full suite of sexual behavior":* Ibid., 4629.

66 *Sex change fish:* J. Godwin, "Neuroendocrinology of Sexual Plasticity in Teleost Fishes," *Frontiers in Neuroendocrinology* 31, no. 2 (2010): 203–16.

66 *Anglerfish with parasitic males:* Theodore W. Pietsch, "Dimorphism, Parasitism, and Sex Revisited: Modes of Reproduction Among Deep-Sea Ceratioid Anglerfishes (Teleostei: Lophiiformes)," *Ichthyological Research* 52, no. 3 (2005): 207–36.

66 *Reproduction in sea horses:* Kai N. Stölting and Anthony B. Wilson, "Male Pregnancy in Seahorses and Pipefish: Beyond the Mammalian Model," *Bioessays* 29, no. 9 (2007): 884–96; Camilla M. Whittington, K. Musolf, S. Sommer, and A. B. Wilson, "Behavioural Cues of Reproductive Status in Seahorses *Hippocampus abdominalis*," *Journal of Fish Biology* 83, no. 1 (2013): 220–26. A developmental gene that enables males in a closely related pipefish to have pregnancy pouches was found by April Harlin-Cognato and her colleagues: A. Harlin-Cognato, E. A. Hoffman, and A. G. Jones, "Gene Cooption Without Duplication During the Evolution of a Male-Pregnancy Gene in Pipefish," *Proceedings of the National Academy of Sciences USA* 103, no. 51 (2006): 19407–12.

67 *a cloud of minute seahorses:* For a video of seahorse mating, see http://vimeo.com/76477081; seahorse males giving birth can be seen at https://www.youtube.com/watch?v=fDfRIIM5iHQ, at https://www.youtube.com/watch?v=2eIuXxp0sxM, and at https://www.youtube.com/watch?v=BPlnqckOPdY; all accessed Sept. 14, 2014.

## Chapter 3: Picky Females, Easy Males

69 *Darwin quotations on sexual selection:* Darwin, *Origin*, 97–99. For those consulting other editions than the one cited above, the quotes occur a few pages into chapter 4.

71 *But tactless wording aside, was Darwin right?:* For a dissenting view, see Joan Roughgarden's *Evolution's Rainbow: Diversity, Gender, and Sexuality in Nature and People* (Berkeley: University of California

Press, 2013). Roughgarden is a distinguished evolutionary ecologist who makes a frontal (and, in the consensus view, greatly exaggerated) assault on the theory of sexual selection as presented by Darwin and greatly improved in the past half century. The book should be read with reviews by two other leading evolutionary biologists, Sarah Blaffer Hrdy (*Nature* 429 [2004]: 19–21) and Alison Jolly (*Science* 304 [2004]: 965–66). Hrdy cautions, "Competition between those of one sex for reproductive access to the other remains a robust explanatory framework, even though it is not the whole story" (p. 21); Jolly writes, "What Darwinian theory needs is not so much radical revision as a simple expansion to take sexual diversity much more seriously" (p. 966). Both admire, as do I, Roughgarden's call for greater attention to the variety of sexual arrangements in the animal and human worlds; I try to pay that attention here.

71 *Darwin went on to write a book about all this:* Charles Darwin, *The Descent of Man, and Selection in Relation to Sex* (1871; rpt.; New York: Penguin Classics, 2004).

72 *"Where one sex invests considerably more":* R. L. Trivers, "Parental Investment and Sexual Selection," in *Sexual Selection and the Descent of Man*, 1871–1971, ed. B. Campbell (Chicago: Aldine, 1972), 173.

72 *consider the cassowary:* Olivia Judson, "Big Bird," *National Geographic* 224, no. 3 (2013): 60–77. For more scientific detail, see L. A. Moore, "Population Ecology of the Southern Cassowary *Casuarius casuarius johnsonii*, Mission Beach, North Queensland," *Journal of Ornithology* 148, no. 3 (2007): 357–66. For a superb documentary, see David Attenborough's *Cassowaries,* produced by Bianca Keeley for the BBC Natural World series.

73 *"His chicks . . . make funny whistling-peeping sounds":* Judson, "Big Bird," 67–68.

73 *"I'm coming back as a female cassowary":* Ibid., 67.

73 *Jacanas:* Stephen T. Emlen and Peter H. Wrege, "Division of Labour in Parental Care Behaviour of a Sex-Role-Reversed Shorebird, the Wattled Jacana," *Animal Behaviour* 68, no. 4 (2004): 847–55; see also the same authors' "Size Dimorphism, Intrasexual Competition, and Sexual Selection in Wattled Jacana (*Jacana jacana*), a Sex-Role-Reversed Shorebird in Panama," *Auk* 121, no. 2 (2004): 391. A dramatic video of jacana reproductive life was available on Sept. 12, 2014, at https://www.youtube.com/watch?v=OitI6o076rY. It's

an excerpt from episode 12 of *The Trials of Life,* narrated by David Attenborough, BBC, 1990.

74 *Jacana female an egg-making machine:* Emlen quoted by the Evolution Library, PBS, http://www.pbs.org/wgbh/evolution/library/01/6/l_016_04.html, accessed Sept. 12, 2014.

74 *Eight thousand species of pair-bonding birds:* A. Cockburn, "Prevalence of Different Modes of Parental Care in Birds," *Proceedings of the Royal Society B: Biological Sciences* 273, no. 1592 (2006): 1375–83.

77 *peahens lay more eggs for males with larger trains:* M. Petrie, T. Halliday, and C. Sanders, "Peahens Prefer Peacocks with Elaborate Trains," *Animal Behavior* 41 (1991): 323–31; Marion Petrie, "Improved Growth and Survival of Offspring of Peacocks with More Elaborate Trains," *Nature* 371, no. 6498 (1994): 598–99; and M. Petrie, P. Cotgreave, and T. W. Pike, "Variation in the Peacock's Train Shows a Genetic Component," *Genetica* 135, no. 1 (2009): 7–11.

78 *Zahavi's "handicap principle":* A. Zahavi, "Mate Selection: A Selection for a Handicap," *Journal of Theoretical Biology* 53 (1975): 205–14; Amotz Zahavi, *The Handicap Principle: A Missing Piece of Darwin's Puzzle* (New York: Oxford University Press, 1999).

78 *Fisher's "runaway selection" theory:* Ronald A. Fisher, *The Genetical Theory of Natural Selection* (1930; repr., New York: Dover, 1958).

78 *Peahens' preference for elaborate peacock tails:* Roslyn Dakin and Robert Montgomerie, "Eye for an Eyespot: How Iridescent Plumage Ocelli Influence Peacock Mating Success," *Behavioral Ecology* 24, no. 5 (2013): 1048–57.

78 *Tail length in male barn swallows:* A. P. Møller, A. Barbosa, J. J. Cuervo, F. de Lope, S. Merino, and N. Saino, "Sexual Selection and Tail Streamers in the Barn Swallow," *Proceedings of the Royal Society: Biological Sciences* 265 (1998): 409–14; and Y. Vortman, A. Lotem, R. Dor, I. J. Lovette, and R. J. Safran, "The Sexual Signals of the East-Mediterranean Barn Swallow: A Different Swallow Tale," *Behavioral Ecology* 22, no. 6 (2011): 1344–52.

79 *Palmate newt filament length:* Jérémie H. Cornuau, Margaux Rat, Dirk S. Schmeller, and Adeline Loyau, "Multiple Signals in the Palmate Newt: Ornaments Help When Courting," *Behavioral Ecology and Sociobiology* 66, no. 7 (2012): 1045–55.

79 *The Trinidadian guppy:* Kenji Karino, Miho Kobayashi, and

Kazuhiro Orita, "Adaptive Offspring Sex Ratio Depends on Male Tail Length in the Guppy," *Ethology* 112, no. 11 (2006): 1050–55.

79 *In the mandrill:* Joanna M. Setchell, "Do Female Mandrills Prefer Brightly Colored Males?" *International Journal of Primatology* 26, no. 4 (2005): 715–35.

79 *even in our relatively pair-bonding species:* Alan F. Dixson, Barnaby J. Dixson, and Matthew Anderson, "Sexual Selection and the Evolution of Visually Conspicuous Sexually Dimorphic Traits in Male Monkeys, Apes, and Human Beings," *Annual Review of Sex Research* 16 (2005): 1–19.

79 *Roseate tern female streamers:* Brian G. Palestis, Ian C. T. Nisbet, Jeremy J. Hatch, Jennifer M. Arnold, and Patricia Szczys, "Tail Length and Sexual Selection in a Monogamous, Monomorphic Species, the Roseate Tern *Sterna dougallii*," *Journal of Ornithology* 153, no. 4 (2013): 1153–63.

79 *Courtship feeding:* David Lack, "Courtship Feeding in Birds," *Auk* 57, no. 2 (1940): 169–78; I. Galvan and J. J. Sanz, "Mate-Feeding Has Evolved as a Compensatory Energetic Strategy That Affects Breeding Success in Birds," *Behavioral Ecology* 22, no. 5 (2011): 1088–95.

80 *Ringdove courtship:* Daniel S. Lehrman, "The Reproductive Behavior of Ring Doves," *Scientific American* 211, no. 5 (1964): 48–54; Kristen E. Mantei, S. Ramakrishnan, P. J. Sharp, and J. D. Buntin, "Courtship Interactions Stimulate Rapid Changes in GnRH Synthesis in Male Ring Doves," *Hormones and Behavior* 54, no. 5 (2008): 669–75. A brief video, accessed Sept. 12, 2014, can be seen here: https://www.youtube.com/watch?v=kOoSDWVM8f8.

81 *Rat courtship:* S. A. Barnett, *The Rat: A Study in Behavior* (Chicago: University of Chicago Press, 1975).

81 *"Sex with Knockout Models":* Emilie F. Rissman, Scott R. Wersinger, Heather N. Fugger, and Thomas C. Foster, "Sex with Knockout Models: Behavioral Studies of Estrogen Receptor Alpha," *Brain Research* 835, no. 1 (1999): 80–90.

82 *Black grouse leks:* Anni Hämäläinen, Rauno V. Alatalo, Christophe Lebigre, Heli Siitari, and Carl D. Soulsbury, "Fighting Behaviour as a Correlate of Male Mating Success in Black Grouse *Tetrao tetrix*," *Behavioral Ecology and Sociobiology* 66, no. 12 (2102): 1577–86.

82 *Uganda kob:* Andrew Balmford, "Social Dispersion and Lekking in Uganda Kob," *Behaviour* 120, no. 3–4 (1992): 177–91.

83 *Consider the majestic red deer:* T. H. Clutton-Brock, F. E. Guinness, and S. D. Albon, *Red Deer: Behavior and Ecology of Two Sexes* (Chicago: University of Chicago Press, 1982). A video of this herd during the rutting season can be seen here: https://www.youtube.com/watch?v=SZCwezLt7Ks, accessed Sept. 12, 2014.

83 *the more basso the croak:* David Lesbarrères, Juha Merilä, and Thierry Lodé, "Male Breeding Success Is Predicted by Call Frequency in a Territorial Species, the Agile Frog (*Rana dalmatina*)," *Canadian Journal of Zoology* 86, no. 11 (2008): 1273–79.

84 *Elephant seals:* B. J. Le Boeuf, "Male-Male Competition and Reproductive Success in Elephant Seals," *American Zoologist* 14 (1974): 163–76. For a video, see http://channel.nationalgeographic.com/channel/videos/elephant-seal-wars/, "Race to Survive: Elephant Seal Wars," National Geographic Channel, accessed Sept. 12, 2014.

84 *4 percent of the males:* B. Le Boeuf and R. Peterson, "Social Status and Mating Activity in Elephant Seals," *Science* 163 (1969): 91–93.

85 *Mammals with larger females:* Katherine Ralls, "Mammals in Which Females Are Larger Than Males," *Quarterly Review of Biology* 51 (1976): 245–76.

86 *"Once female mammals became committed":* Ibid., 262.

87 *Platypus milk proteins:* Wesley C. Warren and 101 other authors, "Genome Analysis of the Platypus Reveals Unique Signatures of Evolution," *Nature* 453, no. 7192 (2008): 175–83.

87 *Oliver Krüger's agility comparison:* Oliver Krüger, "The Evolution of Reversed Sexual Size Dimorphism in Hawks, Falcons and Owls: A Comparative Study," *Evolutionary Ecology* 19, no. 5 (2005): 467–86.

87 *Bringing Ralls up to date on mammals:* P. Lindenfors and B. S. Tullberg, "Evolutionary Aspects of Aggression: The Importance of Sexual Selection," *Advances in Genetics* 75 (2011): 7–22.

88 *Holekamp's comprehensive 2011 review:* K. E. Holekamp, J. E. Smith, C. C. Strelioff, R. C. Van Horn, and H. E. Watts, "Society, Demography and Genetic Structure in the Spotted Hyena," *Molecular Ecology* 21, no. 3 (2012): 613–32.

88 *"Adult females are larger":* Ibid., 614.

89 *"Assumption of the receptive stance":* Micaela Szykman, Russell C. Van Horn, Anne L. Engh, Erin E. Boydston, and Kay E. Holekamp, "Courtship and Mating in Free-Living Spotted Hyenas," *Behaviour* 144, no. 7 (2007): 828.

89 *"There is no external vagina":* Ibid., 816.

90 *Consider the small tropical fish:* M. E. Cummings, "Looking for Sexual Selection in the Female Brain," *Philosophical Transactions of the Royal Society B: Biological Sciences* 367, no. 1600 (2012): 2348–56.

92 *The importance of oxytocin:* C. S. Carter, "Oxytocin Pathways and the Evolution of Human Behavior," *Annual Review of Psychology* 65 (2014): 17–39.

92 *Male fidelity and fatherhood:* T. R. Insel, "The Challenge of Translation in Social Neuroscience: A Review of Oxytocin, Vasopressin, and Affiliative Behavior," *Neuron* 65, no. 6 (2010): 768–79.

92 *How the gene controls receptors in the emotional brain:* Zoe R. Donaldson and Larry J. Young, "The Relative Contribution of Proximal 5' Flanking Sequence and Microsatellite Variation on Brain Vasopressin la Receptor (Avprla) Gene Expression and Behavior," *PLoS Genetics* 9, no. 8 (2013): e1003729.

93 *Remade males stay by females:* Larry J. Young, Roger Nilsen, Katrina G. Waymire, Grant R. MacGregor, and Thomas R. Insel, "Increased Affiliative Response to Vasopressin in Mice Expressing the V1a Receptor from a Monogamous Vole," *Nature* 400, no. 6746 (2004): 766–68.

93 *Miranda Lim and voles:* Miranda M. Lim, Zuoxin Wang, Daniel E. Olazabal, Xianghui Ren, Ernest F. Terwilliger, and Larry J. Young, "Enhanced Partner Preference in a Promiscuous Species by Manipulating the Expression of a Single Gene," *Nature* 429, no. 6993 (2004): 754–57.

93 *Donaldson is exploring the same gene in primates:* Z. R. Donaldson, F. A. Kondrashov, A. Putnam, Y. Bai, T. L. Stoinski, E. A. Hammock, and L. J. Young, "Evolution of a Behavior-Linked Microsatellite-Containing Element in the 5' Flanking Region of the Primate AVPR1A Gene," *BMC Evolutionary Biology* 8 (2008): 180.

93 *Chimpanzee males more dominant and less stable:* W. D. Hopkins, Z. R. Donaldson, and L. J. Young, "A Polymorphic Indel Containing the RS3 Microsatellite in the 5' Flanking Region of the Vasopressin V1a Receptor Gene Is Associated with Chimpanzee (*Pan troglodytes*) Personality," *Genes, Brain, and Behavior* 11, no. 5 (2012): 552–58.

93 *the first research on the gene in wild chimpanzees:* Stephanie F. Anestis, Timothy H. Webster, Jason M. Kamilar, M. Babette Fontenot, David

P. Watts, and Brenda J. Bradley, "AVPR1A Variation in Chimpanzees (*Pan troglodytes*): Population Differences and Association with Behavioral Style," *International Journal of Primatology* 35, no. 1 (2014): 305–24.

94 *Gene linked to partner loyalty in humans:* H. Walum and ten other authors, "Genetic Variation in the Vasopressin Receptor 1a Gene (AVPR1A) Associates with Pair-Bonding Behavior in Humans," *Proceedings of the National Academy of Sciences USA* 105, no. 37 (2008): 14153–56.

## Chapter 4: Primate Possibilities

95 *Lemurs:* Some of the first insights came from the pioneering study by Alison Jolly, *Lemur Behavior: A Madagascar Field Study* (Chicago: University of Chicago Press, 1966). See also her obituary in the *Economist,* accessed Sept. 13, 2014, at http://www.economist.com/news/obituary/21597862-alison-jolly-primatologist-died-february-6th-aged-76-alison-jolly.

96 *Blue-eyed black lemurs:* L. Digby, and A. McLean Stevens, "Maintenance of Female Dominance in Blue-Eyed Black Lemurs (*Eulemur macaco flavifrons*) and Gray Bamboo Lemurs (*Hapalemur griseus griseus*) Under Semi-Free-Ranging and Captive Conditions," *Zoo Biology* 26, no. 5 (2007): 345–61.

96 *Red-bellied vs. crowned lemurs:* B. Marolf, A. G. McElligott, and A. E. Muller, "Female Social Dominance in Two Eulemur Species with Different Social Organizations," *Zoo Biology* 26, no. 3 (2007): 201–14.

96 *Ring-tailed lemurs:* C. M. Drea and A. Weil, "External Genital Morphology of the Ring-Tailed Lemur (*Lemur catta*): Females Are Naturally 'Masculinized,'" *Journal of Morphology* 269, no. 4 (2008): 451–63.

97 *"Sex in the Dark":* Manfred Eberle and Peter M. Kappeler, "Sex in the Dark: Determinants and Consequences of Mixed Male Mating Tactics in *Microcebus murinus*, a Small Solitary Nocturnal Primate," *Behavioral Ecology and Sociobiology* 57, no. 1 (2004): 77–90.

97 *females prefer males who defeat other males:* D. Gomez, E. Huchard, P. Y. Henry, and M. Perret, "Mutual Mate Choice in a Female-Dominant and Sexually Monomorphic Primate," *American Journal of Physical Anthropology* 147, no. 3 (2012): 370–79.

98 *"Mouse lemur females exert tight control":* E. Huchard, C. I. Canale, C. Le Gros, M. Perret, P. Y. Henry, and P. M. Kappeler, "Convenience Polyandry or Convenience Polygyny? Costly Sex Under Female Control in a Promiscuous Primate," *Proceedings of the Royal Society B: Biological Sciences* 279, no. 1732 (2012): 1371.

98 *a book intriguigingly called . . .:* Sarah Blaffer Hrdy, *The Woman That Never Evolved* (Cambridge, MA: Harvard University Press, 1981).

99 *a landmark study of langurs:* S. B. Hrdy, *The Langurs of Abu: Female and Male Strategies of Reproduction* (Cambridge, MA: Harvard University Press, 1977).

100 *"The Optimal Number of Fathers":* S. B. Hrdy, "The Optimal Number of Fathers: Evolution, Demography, and History in the Shaping of Female Mate Preferences," *Annals of the New York Academy of Sciences* 907 (2000): 75–96.

100 *marmosets give birth to twins routinely:* R. A. Harris and seven other authors, "Evolutionary Genetics and Implications of Small Size and Twinning in Callitrichine Primates," *Proceedings of the National Academy of Sciences USA* 111, no. 4 (2014): 1467–72; T. E. Ziegler, S. L. Prudom, and S. R. Zahed, "Variations in Male Parenting Behavior and Physiology in the Common Marmoset," *American Journal of Human Biology* 21, no. 6 (2009): 739–44.

101 *Helpers who refrain from reproduction:* Jeffrey E. Fite, Kimberly J. Patera, Jeffrey A. French, Michael Rukstalis, Elizabeth C. Hopkins, and Corinna N. Ross, "Opportunistic Mothers: Female Marmosets (*Callithrix kuhlii*) Reduce Their Investment in Offspring When They Have to, and When They Can," *Journal of Human Evolution* 49, no. 1 (2005): 122–42.

101 *Consider a pair of golden-white tassel-ear marmosets:* The description is based on the film *Gremlins: Faces in the Forest,* a 1998 co-production of Thirteen/WNET and Survival Anglia, part of the Nature Video Library. It was produced, written, and directed by Nick Gordon; the executive editor and host was George Page; and the scientific adviser was Dr. Marc van Roosmalen.

103 *Baboon males ruled . . . through alliances:* Irven DeVore, "Male Dominance and Mating Behavior in Baboons," in *Sexual Behavior,* ed. Frank Beach (New York: John Wiley, 1965), 266–89. See also a book edited by DeVore: *Primate Behavior* (New York: Holt, Rinehart, and Winston, 1965).

103 *the mother-infant relationship:* I. DeVore, "Mother-Infant Relations in Free-Ranging Baboons," in *Maternal Behavior in Mammals*, ed. H. L. Rheingold (New York: John Wiley, 1963), 305–35.

104 *Smuts's remarkable book:* B. Smuts, *Sex and Friendship in Baboons* (New York, Aldine, 1985).

104 *"agonistic buffering":* A good place to start is Stefanie Henkel, Michael Heistermann, and Julia Fischer, "Infants as Costly Social Tools in Male Barbary Macaque Networks," *Animal Behaviour* 79, no. 6 (2010): 1199–1204.

105 *a paper on male sexual force in primates:* B. B. Smuts and R. W. Smuts, "Male-Aggression and Sexual Coercion of Females in Nonhuman-Primates and Other Mammals: Evidence and Theoretical Implications," *Advances in the Study of Behavior* 22 (1993): 1–63.

105 *Research gathered in a 2009 book:* Martin N. Muller and Richard W. Wrangham, eds., *Sexual Coercion in Primates and Humans: An Evolutionary Perspective on Male Aggression Against Females* (Cambridge, MA: Harvard University Press, 2009). Chapters include Cheryl Knott on orangutans, Martha Robbins on gorillas, Dawn Kitchen and her colleagues on chacma baboons, and Larissa Swedell and Amy Schreier on hamadryas baboons.

106 *Spider monkey male "crashing through the canopy":* K. N. Gibson, L. G. Vick, A. C. Palma, F. M. Carrasco, D. Taub, and G. Ramos-Fernandez, "Intra-community Infanticide and Forced Copulation in Spider Monkeys: A Multi-site Comparison Between Cocha Cashu, Peru, and Punta Laguna, Mexico," *American Journal of Primatology* 70, no. 5 (2008): 485–89. This and subsequent quotes in the passage are from p. 486.

106 *over two years watching wild Sumatran orangs:* ElizaBeth Fox, "Female Tactics to Reduce Sexual Harassment in the Sumatran Orangutan (*Pongo pygmaeus abelii*)," *Behavioral Ecology and Sociobiology* 52, no. 2 (2002): 93–101.

107 *When fertile, orang females actively sought mature males:* C. D. Knott, M. Emery Thompson, R. M. Stumpf, and M. H. McIntyre, "Female Reproductive Strategies in Orangutans, Evidence for Female Choice and Counterstrategies to Infanticide in a Species with Frequent Sexual Coercion," *Proceedings of the Royal Society B: Biological Sciences* 277, no. 1678 (2020): 105–13.

107 *A rare case of sexual bullying in ring-tailed lemurs:* J. A. Parga and A.

R. Henry, "Male Aggression During Mating: Evidence for Sexual Coercion in a Female Dominant Primate?" *American Journal of Primatology* 70, no. 12 (2008): 1187–90.

108 *four out of five Ph.D.s in primatology will go to women:* This estimate was reported in 2001 by Londa Schiebinger in *Has Feminism Changed Science?* (Cambridge, MA: Harvard University Press, 2001), 91.

108 *Jane Goodall's pristine chimpanzees:* Jane van Lawick Goodall, *In the Shadow of Man* (New York, Dell, 1971).

109 *Goodall's scientific masterwork:* Jane Goodall, *The Chimpanzees of Gombe: Patterns of Behavior* (Cambridge, MA: Harvard University Press, 1986). See also her article "Infant Killing and Cannibalism in Free-Living Chimpanzees," *Folia Primatologica* 28, no. 4 (1977): 259–89.

109 *John Mitani's review:* John C. Mitani, "Cooperation and Competition in Chimpanzees: Current Understanding and Future Challenges," *Evolutionary Anthropology: Issues, News, and Reviews* 18, no. 5 (2009): 215–27. See also M. L. Wilson and 29 other authors, "Lethal Aggression in *Pan* Is Better Explained by Adaptive Strategies Than Human Impacts," *Nature* 513, no. 7518 (2014): 414–17.

110 *Red colobus hunting by chimps:* Geza Teleki, *The Predatory Behavior of Wild Chimpanzees* (Lewisburg, PA: Bucknell University Press, 1973); Craig B. Stanford, *The Hunting Apes: Meat Eating and the Origins of Human Behavior* (Princeton, NJ: Princeton University Press, 1999), 118.

110 *Tolerated scrounging and human sharing:* N. G. Blurton Jones, "Tolerated Theft: Suggestions About the Ecology and Evolution of Sharing, Hoarding and Scrounging," *Social Science Information sur les sciences sociales* 26, no. 1 (1987): 31–54.

110 *Meat for sex in a complex ape economy:* Cristina M. Gomes and Christophe Boesch, "Reciprocity and Trades in Wild West African Chimpanzees," *Behavioral Ecology and Sociobiology* 65, no. 11 (2011): 2183–96.

111 *Taï Forest females exercise choice:* R. M. Stumpf and C. Boesch, "The Efficacy of Female Choice in Chimpanzees of the Taï Forest, Côte d'Ivoire," *Behavioral Ecology and Sociobiology* 60, no. 6 (2006): 749–65.

112 *The bonobo genome was sequenced:* Kay Prüfer and forty other authors, "The Bonobo Genome Compared with the Chimpanzee and Human Genomes," *Nature* 486, no. 7404 (2012): 527–31.

112 *As Adrienne Zihlman showed decades ago:* Adrienne L. Zihlman, J. E. Cronin, D. L. Cramer, and Vincent M. Sarich, "Pygmy Chimpanzee as a Possible Prototype for the Common Ancestor of Humans, Chimpanzees, and Gorillas," *Nature* 275 (1978): 744–46.

112 *Bonobos "make love, not war":* For an excellent documentary, see *The Last Great Ape,* produced and directed by Steve Greenwood for the BBC (2006) and then for NOVA/PBS/WGBH Educational Foundation (2007). For information, see http://www.pbs.org/wgbh/nova/bonobos/. For scientific background, see Takayoshi Kano, *The Last Ape: Pygmy Chimpanzee Behavior and Ecology* (Stanford, CA: Stanford University Press, 1992), and T. Furuichi, "Female Contributions to the Peaceful Nature of Bonobo Society," *Evolutionary Anthropology* 20, no. 4 (2011): 131–42.

114 *Hohmann and Fruth studied conflicts related to mating:* Gottfried Hohmann and Barbara Fruth, "Aggression by Bonobos in the Context of Mating," *Behaviour* 140 (2003): 1389–1413.

115 *a 2012 hormonal study of wild bonobos:* Martin Subeck, Tobias Deschner, Grit Schubert, Anja Weltring, and Gottfried Hohmann, "Mate Competition, Testosterone and Intersexual Relationships in Bonobos, *Pan paniscus,*" *Animal Behaviour* 83, no. 3 (2012): 659–69.

115 *sex between females*: Z. Clay, S. Pika, T. Gruber, and K. Zuberbuhler, "Female Bonobos Use Copulation Calls as Social Signals," *Biology Letters* 7, no. 4 (2011): 513–16; and Z. Clay, and K. Zuberbuhler, "Communication During Sex Among Female Bonobos: Effects of Dominance, Solicitation and Audience," *Scientific Reports* 2 (2012): 291.

116 *The two species differ early in development:* V. Wobber, R. Wrangham, and B. Hare, "Bonobos Exhibit Delayed Development of Social Behavior and Cognition Relative to Chimpanzees," *Current Biology* 20, no. 3 (2010): 226–30; V. Woods and B. Hare, "Bonobo but Not Chimpanzee Infants Use Socio-sexual Contact with Peers," *Primates* 52, no. 2 (2011): 111–16; and Elisabetta Palagi and Giada Cordoni, "The Right Time to Happen: Play Developmental Divergence in the Two Pan Species," *PLoS One* 7, no. 12 (2012): e52767.

117 *Bonobo brains more adapted for social cognition and impulse control:* James K. Rilling, Jan Scholz, Todd M. Preuss, Matthew F. Glasser, Bhargav K. Errangi, and Timothy E. Behrens, "Differences Between Chimpanzees and Bonobos in Neural Systems Supporting Social

Cognition," *Social Cognitive and Affective Neuroscience* 7, no. 4 (2012): 369–79.

## Chapter 5: Equal Origins?

119 *"In the beginning there was nature":* Camille Paglia, *Sexual Personae: Art and Decadence from Nefertiti to Emily Dickinson* (New Haven, CT: Yale University Press, 1990), 1.

123 *"Although strong dimorphism is consistently":* J. M. Plavcan, "Sexual Size Dimorphism, Canine Dimorphism, and Male-Male Competition in Primates: Where Do Humans Fit In?" *Human Nature* 23, no. 1 (2012): 52.

125 *lap dancers earn more money at midcycle:* For a review of studies of this and other evidence of changes in sexual attractiveness at mid-cycle, see M. G. Haselton and K. Gildersleeve, "Can Men Detect Ovulation?" *Current Directions in Psychological Science* 20, no. 2 (2011): 87–92. For an elegant study using digitized body movements, see Bernhard Fink, Nadine Hugill, and Benjamin P. Lange, "Women's Body Movements Are a Potential Cue to Ovulation," *Personality and Individual Differences* 53, no. 6 (2012): 759–63.

125 *The "red dress effect" in winter:* On cold days, women wearing red or pink are almost twice as likely to be in the fertile part of the cycle: J. L. Tracy and A. T. Beall, "The Impact of Weather on Women's Tendency to Wear Red or Pink When at High Risk for Conception," *PLoS One* 9, no. 2 (2014): e88852. See also A. T. Beall, and J. L. Tracy, "Women Are More Likely to Wear Red or Pink at Peak Fertility," *Psychological Science* 24, no. 9 (2013): 1837–41.

125 *Polygyny commonly allowed:* G. Murdock and D. White, "Standard Cross-Cultural Sample," *Ethnology* 8 (1969): 329–69; Melvin Ember, Carol R. Ember, and Bobbi S. Low, "Comparing Explanations of Polygyny," *Cross-Cultural Research* 41, no. 4 (2007): 428–40.

126 *Father-infant proximity and mating systems:* Frank W. Marlowe, "Paternal Investment and the Human Mating System," *Behavioural Processes* 51 (2000): 45–61.

126 *"Nonclassical" polyandry throughout the world:* K. E. Starkweather and R. Hames, "A Survey of Non-classical Polyandry," *Human Nature* 23, no. 2 (2012): 149–72.

127 *the custom of* partible paternity*:* Robert S. Walker, Mark V. Flinn, and

Kim R. Hill, "Evolutionary History of Partible Paternity in Lowland South America," *Proceedings of the National Academy of Sciences* 107, no. 45 (2010): 19195–200.

127 *Hunter-gatherers allow polygyny but not much:* Robert S. Walker, Kim R. Hill, Mark V. Flinn, and Ryan M. Ellsworth, "Evolutionary History of Hunter-Gatherer Marriage Practices," *PLoS One* 6, no. 4 (2011): e9066.

127 *Why did those old men get all the women?:* For a discussion, see R. B. Lee and I. DeVore, *Man the Hunter* (Chicago: Aldine, 1968).

128 *"a large number of studies":* Kelly Gildersleeve, Lisa DeBruine, Martie G. Haselton, David A. Frederick, Ian S. Penton-Voak, Benedict C. Jones, and David I. Perrett, "Shifts in Women's Mate Preferences Across the Ovulatory Cycle: A Critique of Harris (2011) and Harris (2012)," *Sex Roles* 69, no. 9–10 (2013): 516–24.

129 Man the Hunter *and* Woman the Gatherer*:* Lee and DeVore, 1968, cited above; Frances Dahlberg, ed., *Woman the Gatherer* (New Haven, CT: Yale University Press, 1981).

129 *Stone tools in an ancient chimp habitat:* J. Mercader, H. Barton, J. Gillespie, J. Harris, S. Kuhn, R. Tyler, and C. Boesch, "4,300-Year-Old Chimpanzee Sites and the Origins of Percussive Stone Technology," *Proceedings of the National Academy of Sciences of the United States of America* 104, no. 9 (2007): 3043–48.

129 *females use a broader spectrum of tools:* Thibaud Gruber, Zanna Clay, and Klaus Zuberbühler, "A Comparison of Bonobo and Chimpanzee Tool Use: Evidence for a Female Bias in the Pan Lineage," *Animal Behaviour* 80, no. 6 (2010): 1023–33.

130 *Women's contribution on average about half:* Frank W. Marlowe, *The Hadza: Hunter-Gatherers of Tanzania* (Berkeley: University of California Press, 2010), chapter 10; and Frank W. Marlowe, "Hunting and Gathering: The Human Sexual Division of Foraging Labor," *Cross-Cultural Research: The Journal of Comparative Social Science* 41, no. 2 (2007): 170–195. The roughly equal division is true overall for warm-climate hunter-gatherers, the ones most relevant to the evolution of our species.

130 *there should be more intense male competition:* F. W. Marlowe and J. C. Berbesque, "The Human Operational Sex Ratio: Effects of Marriage, Concealed Ovulation, and Menopause on Mate Competition," *Journal of Human Evolution* 63, no. 6 (2012): 834–42.

131 *Hunter-gatherer mortality:* Nancy Howell, *Demography of the Dobe Area !Kung* (New York: Academic Press, 1979).

131 *A pair-bonding species with adaptive flexibility:* Robert J. Quinlan, "Human Pair-Bonds: Evolutionary Functions, Ecological Variation, and Adaptive Development," *Evolutionary Anthropology* 17, no. 5 (2008): 227–38.

131 *"Should a paternally caring male desert":* H. Kokko, H. Klug, and M. D. Jennions, "Unifying Cornerstones of Sexual Selection: Operational Sex Ratio, Bateman Gradient and the Scope for Competitive Investment," *Ecology Letters* 15, no. 11 (2012): 1343.

131 *Hadza men experience conflict but invest in children:* Frank W. Marlowe, "Male Care and Mating Effort Among Hadza Foragers," *Behavioral Biology and Sociobiology* 46 (1999): 57–64; Frank W. Marlowe, "A Critical Period for Provisioning by Hadza Men: Implications for Pair Bonding," *Evolution and Human Behavior* 24 (2003): 217–29.

131 *Male investment in children in other hunter-gatherers:* Frank W. Marlowe, "Paternal Investment and the Human Mating System," *Behavioural Processes* 51 (2000): 45–61; and Michael Gurven and Kim Hill, "Why Do Men Hunt?," *Current Anthropology* 50, no. 1 (2009): 51–74. Male investment is not limited to fathers: Kim Hill and A. Magdalena Hurtado, "Cooperative Breeding in South American Hunter-Gatherers," *Proceedings of the Royal Society: Biological Sciences* 276, no. 1674 (2009): 3863–70.

131 *Grandmothers, menopause, and male show-offs:* Kristen Hawkes, "Grandmothers and the Evolution of Human Longevity," *American Journal of Human Biology* 15, no. 3 (2003): 380–400; and Kristen Hawkes, James F. O'Connell, and James E. Coxworth, "Family Provisioning Is Not the Only Reason Men Hunt: A Comment on Gurven and Hill," *Current Anthropology* 51, no. 2 (2010): 259–64.

131 *Menopause reduces competition between generations of women:* R. Mace, "Cooperation and Conflict Between Women in the Family," *Evolutionary Anthropology* 22, no. 5 (2013): 251–58.

132 *Flow of calories among !Kung:* Nancy Howell, *Life Histories of the Dobe !Kung: Food, Fatness, and Well-Being over the Life Span* (Berkeley: University of California Press, 2010).

132 *only humans have such extensive provisioning:* Jane B. Lancaster and Chet S. Lancaster, "The Watershed: Change in Parental-Investment and Family-Formation Strategies in the Course of Human

Evolution," in *Parenting Across the Life Span: Biosocial Dimensions*, ed. Jane B. Lancaster, Jeanne Altmann, Alice S. Rossi, and Lonnie R. Sherrod (New York: Aldine De Gruyter, 1987), 187–205; updated in an important theoretical article by Hillard Kaplan, Kim Hill, Jane Lancaster, and A. Magdalena Hurtado, "A Theory of Human Life History Evolution: Diet, Intelligence, and Longevity," *Evolutionary Anthropology* 9, no. 4 (2000): 156–85.

132 *We are cooperative breeders:* Sarah Blaffer Hrdy, *Mothers and Others: The Evolutionary Origins of Mutual Understanding* (Cambridge, MA: Harvard University Press, 2009).

132 *confirmed in the past few years:* Karen L. Kramer, "Cooperative Breeding and Its Significance to the Demographic Success of Humans," *Annual Review of Anthropology* 39, no. 1 (2010): 417–36; Howell, *Life Histories of the Dobe !Kung;* B. F. Codding, R. B. Bird, and D. W. Bird, "Provisioning Offspring and Others: Risk-Energy Trade-offs and Gender Differences in Hunter-Gatherer Foraging Strategies," *Proceedings of the Royal Society* 278, no. 1717 (2011): 2502–09; and Courtney L. Meehan, Robert Quinlan, and Courtney D. Malcom, "Cooperative Breeding and Maternal Energy Expenditure Among Aka Foragers," *American Journal of Human Biology* 25, no. 1 (2013): 42–57.

132 *the !Kung San, or Ju/'hoansi:* The exclamation point refers to an implosive consonant, or "click," but ignoring it (as in "Kung") gives a reasonable approximation of the name; likewise, the alternative name can be approximated with "Jutoansi." For the fundamental and classic descriptions of their way of life, see Richard Lee's *The !Kung San: Men, Women and Work in a Foraging Society* (Cambridge: Cambridge University Press, 1979), and Lorna Marshall's two books, *The !Kung of Nyae Nyae* (1976) and *Nyae Nyae !Kung: Beliefs and Rites* (1999), both published by Harvard University Press.

132 *Shostak's studies of !Kung women:* Marjorie Shostak, *Nisa: The Life and Words of a !Kung Woman* (Cambridge, MA: Harvard University Press, 1981); and Marjorie Shostak, *Return to Nisa* (Cambridge, MA: Harvard University Press, 2000).

134 *Men did two-thirds of the talking:* "Politics, Sexual and Non-Sexual, in an Egalitarian Society," in *Politics and History in Band Societies*, ed. Eleanor Leacock and Richard Lee (New York: Cambridge University Press, 1982), 37–59.

135 *Interviewing a group of men about animal behavior:* Nicholas G. Blurton Jones and Melvin J. Konner, "!Kung Knowledge of Animal Behavior," in *Kalahari Hunter-Gatherers: Studies of the !Kung San and Their Neighbors*, ed. Richard B. Lee and Irven DeVore (Cambridge, MA: Harvard University Press, 1976).

135 Women Like Meat: Megan Biesele, *Women Like Meat: The Folklore and Foraging Ideology of the Kalahari Ju/'hoan* (Bloomington: Indiana University Press, 1993).

135 *Trance-dance ritual:* Richard Lee, "The Sociology of !Kung Bushman Trance Performances," in *Trance and Possession States*, ed. R. Prince (Montreal: Bucke Memorial Society, 1968), 35–54; Lorna Marshall, "The Medicine Dance of the !Kung Bushmen," *Africa* 39 (1981): 347–81; and Melvin Konner, "Transcendental Medication," *Sciences* 25, no. 3 (1985): 2–4. For a film of the ritual, see John Marshall's *N/um Tchai: The Ceremonial Dance of the !Kung Bushmen* (Watertown, MA: Documentary Educational Resources, 1969), available for purchase or rental at www.der.org.

137 *"I ran so fast . . .":* Shostak, *Nisa,* 102.

137 *girls had dramatic initiation rites:* Shostak, *Nisa,* 134–35; Marshall, *The Nyae Nyae !Kung: Beliefs and Rites,* chapter 9.

138 *the leveling effect of giving:* Lorna Marshall, "Sharing, Talking, and Giving: Relief of Social Tensions Among the !Kung," in *Kalahari Hunter-Gatherers: Studies of the !Kung San and Their Neighbors*, ed. Richard B. Lee and Irven DeVore (Cambridge, MA: Harvard University Press, 1976), 349–71; and Polly Wiessner, "Leveling the Hunter: Constraints on the Status Quest in Foraging Societies," in *Food and the Status Quest: An Interdisciplinary Perspective*, ed. Polly Wiessner and Wulf Schiefenhövel (Providence, RI: Berghahn, 1996), 171–91.

138 *women should give birth alone:* M. J. Konner and M. J. Shostak, "Timing and Management of Birth Among the !Kung: Biocultural Interaction in Reproductive Adaptation," *Cultural Anthropology* 2 (1987): 11–28; and Megan Biesele, "An Ideal of Unassisted Birth: Hunting, Healing, and Transformation Among the Kalahari Ju/'hoansi," in *Childbirth and Authoritative Knowledge: Cross-Cultural Perspectives*, ed. Robbie E. Davis-Floyd and Carolyn F. Sargent (Berkeley: University of California Press, 1997).

138 *Families never left to their own devices:* Nancy Howell, *Life Histories*

*of the Dobe !Kung: Food, Fatness, and Well-Being over the Life Span* (Berkeley: University of California Press, 2010).

139 *Lee documented twenty-two cases:* Lee, *The !Kung San,* 381–400.

140 *"A woman has to want her husband and her lover equally":* Shostak, *Nisa,* 247.

140 *"Women are strong":* Ibid., 247.

140 *"!Kung women themselves refer to":* Ibid., 220–21.

141 *early in what would become feminist anthropology:* M. Z. Rosaldo and L. Lamphere, eds., *Woman, Culture and Society* (Stanford, CA: Stanford University Press, 1974). For a basic overview of women's roles cross-culturally, see Dorothy Hammond and Alta Jablow's *Women in Cultures of the World* (Menlo Park, CA: Cummings, 1976).

141 *Hunter-gatherers have less assymmetry of power:* Karen L. Endicott, "Gender Relations in Hunter-Gatherer Societies," in *The Cambridge Encyclopedia of Hunters and Gatherers,* ed. Richard B. Lee and Richard Daly (Cambridge: Cambridge University Press, 1999), 411.

142 *Diets average about half animal flesh:* Marlowe, *The Hadza,* chapter 10, and "Hunting and Gathering: The Human Sexual Division of Foraging Labor."

142 *Women involved in hunting:* Rebecca Bliege Bird and Douglas W. Bird, "Why Women Hunt: Risk and Contemporary Foraging in a Western Desert Aboriginal Community," *Current Anthropology* 49, no. 4 (2008): 655–93; Andrew J. Noss and Barry S. Hewlett, "The Contexts of Female Hunting in Central Africa," *American Anthropologist* 103 (2001): 1024–40; Agnes Estioko-Griffin, "Women as Hunters: The Case of an Eastern Cagayan Agta Group," in *The Agta of Northeastern Luzon: Recent Studies,* ed. P. Bion Griffin and Agnes Estioko-Griffin (Cebu City, Philippines: San Carlos, 1985), 18–32. M. J. Goodman, P. B. Griffin, A. A. Estiokogriffin, and J. S. Grove. "The Compatibility of Hunting and Mothering among the Agta Hunter-Gatherers of the Philippines," *Sex Roles* 12, no. 11-1 (1985): 1199–1209.

142 *"What's a Mother to Do?":* Steven L. Kuhn and Mary C. Stiner, "What's a Mother to Do? The Division of Labor Among Neandertals and Modern Humans in Eurasia," *Current Anthropology* 47, no. 6 (2006): 953–81.

143 *Neanderthals ground, cooked, and ate barley:* Amanda G. Henry, Alison S. Brooks, and Dolores R. Piperno, "Microfossils in Calculus Demonstrate Consumption of Plants and Cooked Foods

in Neanderthal Diets (Shanidar III, Iraq; Spy I and II, Belgium)," *Proceedings of the National Academy of Sciences* 108, no. 2 (2011): 486–91.

143 *growing evidence in Africa:* Curtis W. Marean and thirteen other authors, "Early Human Use of Marine Resources and Pigment in South Africa During the Middle Pleistocene," *Nature* 449, no. 7164 (2007): 905–08.

143 *the most caring fathers on record:* Barry S. Hewlett, *Intimate Fathers: The Nature and Context of Aka Pygmy Paternal Infant Care* (Ann Arbor: University of Michigan Press, 1991).

## Chapter 6: Cultivating Dominance

145 *warfare emerged with much greater intensity:* Patricia M. Lambert, "The Archaeology of War: A North American Perspective," *Journal of Archaeological Research* 10, no. 3 (2002): 207–41.

146 *A very similar process in Europe and the Near East:* R. Brian Ferguson, "The Prehistory of War and Peace in Europe and the Near East," in *War, Peace and Human Nature: The Convergence of Evolutionary and Cultural Views*, ed. Douglas P. Fry (New York: Oxford University Press, 2013), 191–240.

146 *One of the earliest shifts to agriculture:* Simone Riehl, Mohsen Zeidi, and Nicholas J. Conard, "Emergence of Agriculture in the Foothills of the Zagros Mountains of Iran," *Science* 341, no. 6141 (2013): 65–67.

147 *"Second Earth":* Marvin Harris, *Culture, People, Nature: An Introduction to General Anthropology*, 7th ed. (New York: Longman, 1997).

148 *worldwide intensification of hunting and gathering:* Melinda A. Zeder, "The Broad Spectrum Revolution at 40: Resource Diversity, Intensification, and an Alternative to Optimal Foraging Explanations," *Journal of Anthropological Archaeology* 31, no. 3 (2012): 241–64.

148 *Northwest Coast Indians:* M. Susan Walter, "Polygyny, Rank, and Resources in Northwest Coast Foraging Societies," *Ethnology* 45, no. 1 (2006): 41–57. See also D. W. Sellen and D. J. Hruschka. "Extracted-Food Resource-Defense Polygyny in Native Western North American Societies at Contact," *Current Anthropology* 45, no. 5 (2004): 707–14.

149 *A spectacular site:* Oliver Dietrich, Manfred Heun, Jens Notroff, Klaus Schmidt, and Martin Zarnkow, "The Role of Cult and Feasting in the Emergence of Neolithic Communities: New Evidence from

Göbekli Tepe, South-Eastern Turkey," *Antiquity* 86, no. 333 (2012): 674–95. For a popular account with photos, see Charles C. Mann, "The Birth of Religion," *National Geographic Magazine,* June 2011.

149 *The shift to agriculture worsened health:* A. Mummert, E. Esche, J. Robinson, and G. J. Armelagos, "Stature and Robusticity During the Agricultural Transition: Evidence from the Bioarchaeological Record," *Economics and Human Biology* 9, no. 3 (2011): 284–301.

150 *"Health Versus Fitness":* Patricia M. Lambert, "Health Versus Fitness," *Current Anthropology* 50, no. 5 (2009): 603–08.

150 *Skeletons of 200 Natufians:* Vered Eshed, Avi Gopher, Ron Pinhasi, and Israel Hershkovitz, "Paleopathology and the Origin of Agriculture in the Levant," *American Journal of Physical Anthropology* 143, no. 1 (2010): 121–33.

151 *Tchambuli women unadorned while men fussed with their hair:* Margaret Mead, *Male and Female: A Study of the Sexes in a Changing World* (New York: Dell, 1968).

151 *Mosuo women kept husbands at a distance:* Cai Hua, *A Society Without Fathers or Husbands: The Na of China* (Brooklyn, New York: Zone Books, 2001), translated by Asti Hustvedt from the 1997 French edition, *Une société sans père ni mari: Les Na de Chine* (Presses Universitaires de France).

152 *Minangkabau women had more influence:* Michael Peletz, "The Exchange of Men in Nineteenth-Century Negeri Sembilan (Malaya)," *American Ethnologist* 14, no. 3 (1987): 449–69; and Jennifer Krier, "The Marital Project: Beyond the Exchange of Men in Minangkabau Marriage," *American Ethnologist* 27, no. 4 (2000): 877–97.

152 *a matrilineal kingdom:* R. S. Rattray, *Ashanti* (Oxford: Clarendon, 1923); and Meyer Fortes, "Kinship and Marriage Among the Ashanti," in *African Systems of Kinship and Marriage*, ed. A. R. Radcliffe-Brown and Daryll Forde (Oxford: Oxford University Press, 1950), 252–85.

152 *In matrilineal societies:* John Hartung, "Matrilineal Inheritance: New Theory and Analysis," *Behavioral and Brain Sciences* 8, no. 4 (1985): 661–70; see also analyses of the function and evolution of these systems by Laura Fortunato, "The Evolution of Matrilineal Kinship Organization," *Proceedings of the Royal Society: Biological Sciences* 279, no. 1749 (2012): 4939–45, and by Constance Holden, Rebecca Sear, and Ruth Mace, "Matriliny as Daughter-Biased Investment," *Evolution and Human Behavior* 24, no. 2 (2003): 99–112.

152 *"deeply suffused with ambivalence":* Michael G. Peletz, *Reason and Passion: Representations of Gender in a Malay Society* (Berkeley: University of California Press, 1996), 311.

152 *"Through a complex chain of symbolic associations":* Ibid., 313.

153 *"Like the ash on a tree trunk":* Ibid., 335.

153 *"the matrilineal puzzle":* Ibid., 337.

153 *"Women predominated in many rituals":* Michael G. Peletz, "Gender Pluralism: Muslim Southeast Asia since Early Modern Times." *Social Research* 78, no. 2 (2011): 659–86, p. 662.

154 *Na ethnography:* Hua, *A Society Without Fathers or Husbands.*

154 *"an extreme case":* Stevan Harrell, "Review of *A Society Without Fathers or Husbands: The Na of China,*" *American Anthropologist* 104, no. 3 (2002): 983.

154 *others have noted:* Eileen Walsh, "Review of *A Society Without Fathers or Husbands: The Na of China,*" *American Ethnologist* 29, no. 4 (2002): 1043–45; Tami Blumenfield, "The Na of Southwest China: Debunking the Myths," 2002, unpublished manuscript, cited by permission; Siobhán Mattison, Brooke Scelza, and Tami Blumenfield, "Paternal Investment and the Positive Effects of Fathers among the Matrilineal Mosuo of Southwest China," *American Anthropologist* 116, no. 3 (2014): 591–610.

154 *"Relations between men and women":* Polly Wiessner, "Alienating the Inalienable: Marriage and Money in a Big Man Society," in *The Scope of Anthropology*: Maurice Godelier's Work in Context, ed. Laurent Dousset and Serge Tcherkezoff (New York: Berghahn Books, 2012), 67–85.

155 *"capital comes into the world":* The German original reads, "Wenn das Geld, nach Augier, 'mit natürlichen Blutflecken auf einer Backe zur Welt kommt,' so das Kapital von Kopf bis Zeh, aus allen Poren, blut- und schmutz-triefend." Karl Marx, *Das Kapital: Kritik der Politischen Oekonomie* (Hamburg: O. Meissner, 1883), 787. Accessed via Google Books on Sept. 13, 2014, http://books.google.com/books?id=xdYDAAAAMAAJ&source=gbs_navlinks_s.

155 *What we call civilization:* The characterization that follows is largely drawn from Bruce G. Trigger, *Understanding Civilizations: A Comparative Study* (New York: Cambridge University Press, 2003). Other works are cited below as needed to reference more specific points.

157 *Women were subjugated in all early civilizations:* Ibid., 142 and chapter 9.

157 *"repetitive, interruptible, non-dangerous":* Judith K. Brown, "A Note on the Division of Labor by Sex," *American Anthropologist* 72, no. 5 (1970): 1077.

158 *A systematic study of 185 societies:* George Peter Murdock and Caterina Provost, "Factors in the Division of Labor by Sex: A Cross-Cultural Analysis," *Ethnology* 12, no. 2 (1973): 203–25.

159 *A distinction between* gemeinschaft *and* gesellschaft: Ferdinand Tönnies, *Community and Society*, trans. C. P. Loomis (New York: Harper & Row, 1957).

160 *bridewealth to present:* For overviews, see Jane Fishburne Collier, *Marriage and Inequality in Classless Societies* (Stanford, CA: Stanford University Press, 1993), and Jack Goody and Stanley J. Tambiah's *Bridewealth and Dowry,* Cambridge Papers in Social Anthropology (New York: Cambridge University Press, 1973). "Bride price" is a misnomer for bridewealth, since it is not a purchase.

160 *Yanomami men who had killed another man:* Napoleon A. Chagnon, "Life Histories, Blood Revenge, and Warfare in a Tribal Population," *Science* 239 (1988): 985–92.

160 *"When a victim is beheaded":* Renato Rosaldo, *Ilongot Headhunting 1883–1974: A Study in Society and History* (Stanford, CA: Stanford University Press, 1980), 140–41.

160 *Homicide has been part of our lives:* For further references on violence in the fossil record, see Melvin Konner, "Human Nature, Ethnic Violence, and War," in *The Psychology of Resolving Global Conflicts: From War to Peace,* vol. 1, ed. Mari Fitzduff and Chris E. Stout (Westport, CT.: Praeger Security International, 2006). See also Lawrence H. Keeley, *War Before Civilization: The Myth of the Peaceful Savage* (New York: Oxford University Press, 1996); Steven LeBlanc and Katherine E. Register, *Constant Battles: The Myth of the Peaceful, Noble Savage* (New York: St. Martin's, 2003); and Steven Pinker, *The Better Angels of Our Nature: Why Violence Has Declined* (New York: Viking, 2011). For an alternative view, see Douglas P. Fry, ed., *War, Peace, and Human Nature: The Convergence of Evolutionary and Cultural Views* (Oxford: Oxford University Press, 2013).

161 *Study by Fry and Söderberg:* D. P. Fry and P. Soderberg, "Lethal Aggression in Mobile Forager Bands and Implications for the Origins of War," *Science* 341, no. 6143 (2013): 270–73.

162 *"interpretive pacifications":* Keeley, *War Before Civilization,* 20.

162 *History as ongoing, expansionist tribal warfare:* See Pinker, *Better Angels,* and Andrew Bard Schmookler, *The Parable of the Tribes: The Problem of Power in Social Evolution* (Berkeley: University of California Press, 1983).

162 *"a sour ferment":* Arnold Toynbee and Edward DeLos Myers, *A Study of History,* vol. 1 (Oxford: Oxford University Press, 1948), 9.

162 *"Sing, O Goddess, the ruinous wrath of Achilles": The Iliad of Homer,* trans. Ennis Rees (Oxford: Oxford University Press, 1991).

163 *"the poem of force":* Simone Weil, *The Iliad; or, the Poem of Force: A Critical Edition* (New York: Peter Lang, 2006).

163 *laid low by his vengeful wife:* Aeschylus, "Agamemnon," in *Aeschylus I*, ed. David Greene and Richmond Lattimore (Chicago: University of Chicago/Modern Library, 1942), 39–101.

163 *Sacrifice of Iphigeneia:* Euripides, *Iphigeneia at Aulis*, trans. W. S. Merwin and George E. Dimock Jr. (New York: Oxford University Press, 1978).

164 *"defiles":* Genesis 34:2. This and subsequent quotes are from the King James version.

164 *"all their wealth . . . and their wives":* Genesis 34:29.

164 *"And the children of Israel took all the women":* Numbers 31:9–10.

164 *"Have ye saved all the women alive?":* Numbers 31:15.

164 *"Now therefore kill":* Numbers 31:17.

164 *"played the whore":* Judges 19:2

165 *"And this is the thing that ye shall do":* Judges 21:11–12.

165 *"Go and lie in wait":* Judges 21:20–21.

165 *"In those days":* Judges 21:25.

165 *David and Bathsheba:* The story is told in 2 Samuel, chapter 11, which in the end does at least tell us that David displeased the Lord.

165 *Great Hindu epic of rivalry and war: Mahabharata,* 35th anniversary ed., trans. William Buck (Berkeley: University of California Press, 2012).

166 *The Bhagavad Gita: The Bhagavad Gita*, trans. Laurie L. Patton (New York: Penguin Classics, 2008).

166 *"The king may enter the harem": Indian History Sourcebook: The Laws of Manu, c. 1500 BCE,* trans. G. Buhler, chapter 7, verses 216 and 221; published by Fordham University at http://www.fordham.edu/halsall/india/manu-full.asp, accessed Sept. 13, 2014. Although the text says that men must honor women, it also makes clear that

they are subject to men's rule. For example, chapter 5, verse 148: "In childhood a female must be subject to her father, in youth to her husband, when her lord is dead to her sons; a woman must never be independent." We will see that similar views were held in the West until quite recently.

166 *Polygyny in various religions:* See the Wikipedia entry "polygamy" at http://en.wikipedia.org/wiki/Polygamy, accessed Sept. 13, 2014.

167 *Despotism and reproduction:* L. L. Betzig, *Despotism and Differential Reproduction: A Darwinian View of History* (New York: Aldine, 1986).

167 *Betzig's 2012 update:* Laura Betzig, "Means, Variances, and Ranges in Reproductive Success: Comparative Evidence," *Evolution and Human Behavior* 33, no. 4 (2012): 309–17.

168 *"lesser known chiefs":* Berys N. Heuer, "Maori Women in Traditional Family and Tribal Life," *Journal of the Polynesian Society* 78, no. 4 (1969): 455.

168 *paramount chief of the Powhatan:* See the Wikipedia entry "Chief Powhatan" for the drawing and references: http://en.wikipedia.org/wiki/Chief_Powhatan, accessed Sept. 13, 2014.

168 *Kamehameha's wives:* Jocelyn Linnekin, *Sacred Queens and Women of Consequence: Rank, Gender, and Colonialism in the Hawaiian Islands* (Ann Arbor: University of Michigan Press, 1990).

168 *six wives and at least twenty-five children:* "Zulu King Zwelithini's Sixth Wife 'Needs Palace,'" BBC News Africa, http://www.bbc.com/news/world-africa-19489196, accessed Sept. 13, 2014.

168 *"Humans, like most other species":* Bobbi S. Low, "Women's Lives There, Here, Then, Now: A Review of Women's Ecological and Demographic Constraints Cross-Culturally," *Evolution and Human Behavior* 26, no. 1 (2005): 66.

169 *"For humans, as for other mammals":* Ibid., 68.

169 *Having co-wives reduces surviving children:* M. Borgerhoff Mulder, "Women's Strategies in Polygynous Marriage: Kipsigis, Datoga, and Other East African Cases, *Human Nature* 3 (1992): 45–70; B. I. Strassmann, "Social Monogamy in a Human Society: Marriage and Reproductive Success Among the Dogon," in *Monogamy: Mating Strategies and Partnerships in Birds, Mammals, and Humans*, ed. Ulrich Reichard and C. Boesch (Cambridge: Cambridge University Press, 2003), 177–89; D. W. Sellen, "Polygyny and Child Growth in a Traditional Pastoral Society: The Case of the Datoga of Tanzania,"

*Human Nature: An Interdisciplinary Biosocial Perspective* 10, no. 4 (1999): 329–71.

169 *"Sex* is *power"*: Paglia, *Sexual Personae,* 23.

170 *"The violent subjugation of women and girls"*: Kathy L. Gaca, "Girls, Women, and the Significance of Sexual Violence in Ancient Warfare," in *Sexual Violence in Conflict Zones*, ed. Elizabeth D. Heineman (Philadelphia: University of Pennsylvania Press, 2011), 87.

170 *"the objective of taking captive"*: Ibid.

170 *"have control over the procreative capabilities"*: David Wyatt, *Slaves and Warriors in Medieval Britain and Ireland, 800–1200* (Leiden: Brill, 2009), 132.

171 *"cut the throat"*: Ibid., 143

171 *The Nuer organization for predatory expansion:* Raymond C. Kelly, *The Nuer Conquest: The Structure and Development of an Expansionist System* (Ann Arbor: University of Michigan Press, 1985).

171 *In the Crusades, rape was standard practice:* Wyatt, *Slaves and Warriors,* 124; and Anne Curry, "The Theory and Practice of Female Immunity in the Medieval West," in *Sexual Violence in Conflict Zones*, ed. Elizabeth D. Heineman (Philadelphia: University of Pennsylvania Press, 2011), 173–88.

171 *Sixteen million descendants of one man:* Tatiana Zerjal and twenty-two other authors, "The Genetic Legacy of the Mongols," *American Journal of Human Genetics* 72, no. 3 (2003): 717–21.

172 *the same proportion of Irish men:* L. T. Moore, B. McEvoy, E. Cape, K. Simms, and D. G. Bradley, "A Y-Chromosome Signature of Hegemony in Gaelic Ireland," *American Journal of Human Genetics* 78, no. 2 (2006): 334–38.

172 *War as runaway sexual selection:* Bobbi S. Low, *Why Sex Matters: A Darwinian Look at Human Behavior* (Princeton, NJ: Princeton University Press, 2000), 217.

## Chapter 7: Samson's Haircut, Achilles' Heel

175 *A classic study of how well women do:* Martin King Whyte, *The Status of Women in Preindustrial Societies* (Princeton, NJ: Princeton University Press, 1978).

175 *"solitary, poor, nasty, brutish, and short"*: Thomas Hobbes and C. B. Macpherson, *Leviathan* (New York: Penguin Classics, 1968), 89.

177 *The take included 11,000 tons of wheat:* Robert B. Coote, *Early Israel: A New Horizon* (Minneapolis: Fortress, 1990), 37.

177 *"The scribe arrives":* Ibid., 5.

179 *Nazis kept Jewish sex slaves:* Sonja Maria Hedgepeth and Rochelle G. Saidel, eds., *Sexual Violence Against Jewish Women During the Holocaust* (Dartmouth, NH: University Press of New England, 2010). Chapters in this book also describe other Nazi sexual crimes against Jewish women.

179 *"comfort women":* Yuma Totani, "Legal Responses to World War II Sexual Violence: The Japanese Experience," in *Sexual Violence in Conflict Zones*, ed. Elizabeth D. Heineman (Philadelphia: University of Pennsylvania Press, 2011), 217–31.

179 *Soviet soldiers raped German women:* Anthony Beevor, *Berlin: The Downfall 1945* (New York: Cambridge University Press, 2007).

179 *"Copulation without conversation":* Osmar White, *Conquerors' Road: An Eyewitness Report of Germany 1945* (New York: Cambridge University Press, 2003), 98.

179 *all occupying armies have left children behind:* For an overview, statistics, and references, see the Wikipedia entry "War children," accessed April 9, 2014, at http://en.wikipedia.org/wiki/War_children#War_children_of_World_War_II. See also Maria Hohn, *GIs and Fräuleins: The German-American Encounter in 1950s West Germany* (Chapel Hill: University of North Carolina Press, 2002).

180 *Pinker's* Better Angels*:* Pinker, *Better Angels of Our Nature.*

181 *"They were supposed to draw the line":* Walter Scheidel "A Peculiar Institution? Greco-Roman Monogamy in Global Context," *History of the Family* 14, no. 3 (2009): 283.

181 *"could marry several in a row":* Ibid., 283.

181 *Jews allowed polygyny:* Harvey E. Goldberg, *Jewish Passages: Cycles of Jewish Life* (Berkeley: University of California Press, 2003), 143–45.

181 *Christian monogamy in a Greco-Roman context:* Scheidel, "A Peculiar Institution?" 289.

183 *A tendency with adaptive flexibility:* See Quinlan, "Human Pair-Bonds," for a thoughtful review and discussion.

183 *The Nayar of India:* Kathleen Gough, "The Nayars and the Definition of Marriage," *Journal of the Royal Anthropological Institute* 89 (1959): 23–34; Melinda A. Moore, "Symbol and Meaning in Nayar Marriage Ritual," *American Ethnologist* 15, no. 2 (1988): 254–73.

184 *Monogamy* independently *arose sixty-one different times:* D. Lukas and T. H. Clutton-Brock, "The Evolution of Social Monogamy in Mammals," *Science* 341, no. 6145 (2013): 526–30.

184 *Opie looked at 230* primate *species:* C. Opie, Q. D. Atkinson, R. I. M. Dunbar, and S. Shultz, "Male Infanticide Leads to Social Monogamy in Primates," *Proceedings of the National Academy of Sciences of the United States of America* 110, no. 33 (2013): 13328–32; see also C. Opie, Q. D. Atkinson, R. I. M. Dunbar, and S. Shultz, "Reply to Dixson: Infanticide Triggers Primate Monogamy," *Proceedings of the National Academy of Sciences of the United States of America* 110, no. 51 (2013): E4938.

184 *The controversy continued:* D. Lukas and T. H. Clutton-Brock, "Evolution of Social Monogamy in Primates Is Not Consistently Associated with Male Infanticide," *Proceedings of the National Academy of Sciences of the United States of America* 111, no. 17 (2014): E1674; C. Opie, Q. D. Atkinson, R. I. M. Dunbar, and S. Shultz. "Reply to Lukas and Clutton-Brock: Infanticide Still Drives Primate Monogamy," *Proceedings of the National Academy of Sciences of the United States of America* 111, no. 17 (2014): E1675.

186 *Fathers paying dowries for grandsons' reproductive success:* M. Dickemann, "Ecology of Mating Systems in Hypergynous Dowry Societies," *Social Science Information sur les sciences sociales* 18, no. 2 (1979): 163–95.

186 *"Graze where you will":* William Shakespeare, *The Yale Shakespeare: The Complete Works,* edited and annotated under the direction of the Department of English, Yale University (New York: Barnes & Noble Books, 1993), *Romeo and Juliet,* act 3, scene 5, p. 925.

186 *Falling in love a "discourse of defiance":* Lila Abu-Lughod, *Veiled Sentiments: Honor and Poetry in a Bedouin Society* (Berkeley: University of California Press, 1986), xix.

187 *Women in ancient Greece and Rome:* This discussion is based on Sarah B. Pomeroy's *Goddesses, Whores, Wives, and Slaves: Women in Classical Antiquity* (New York: Schocken, 1995). See also the Wikipedia entry "Women's Rights" at http://en.wikipedia.org/wiki/Women%27s_rights, accessed Sept. 13, 2014.

187 *The drama of Jane Austen:* Jane Austen, *Pride and Prejudice* (New York: Knopf/Everyman's Library, 1991); and Jane Austen, *Emma* (New York: Modern Library Paperback, 2001).

188 *"And by the way in the new Code of Laws":* Miriam Schneir, *Feminism: The Essential Historical Writings* (New York: Vintage, 1994), 3.

189 *de Gouges's* Déclaration des droits de la femme et de la citoyenne*:* The full text in English was accessed Sept. 13, 2014, at http://chnm.gmu.edu/revolution/d/293/. The French text was available on the same date at http://fr.wikisource.org/wiki/D%C3%A9claration_des_droits_de_la_femme_et_de_la_citoyenne.

189 *Wollstonecraft's* Vindication*:* Mary Wollstonecraft, *A Vindication of the Rights of Woman* (Dover Thrift, 2012), Kindle edition.

189 *Willard's "Female Education":* Emma Willard, "Mrs. Willard's Plan for Female Education: An Address to the Public; Particularly to the Members of the Legislature of New-York, Proposing a Plan for Improving Female Education," accessed Sept. 13, 2014, at http://www.emmawillard.org/archive/mrs-willards-plan-education, on the website of the Emma Willard School, still in Troy, New York.

190 *the first women's rights convention:* Judith Wellman, "The Seneca Falls Women's Rights Convention: A Study of Social Networks," *Journal of Women's History* 3, no. 1 (1991): 9–37.

190 *"We are continually told that civilization":* John Stuart Mill, *The Basic Writings of John Stuart Mill: "On Liberty," "The Subjection of Women," and "Utilitarianism"* (New York: Random House, 2010), 154. For other editions, the quote is near the beginning of chapter 2 of "The Subjection of Women."

190 *Very slowly, men were won over:* For further discussion and references for this and the next few paragraphs, see the Wikipedia entry "Women's rights," at http://en.wikipedia.org/wiki/Women's_rights, accessed Sept. 13, 2014.

191 *"the right to copulate with a feeling of security":* See Jill Lepore's "Birth-right: What's Next for Planned Parenthood?" *New Yorker,* November 14, 2011, at http://www.newyorker.com/reporting/2011/11/14/111114fa_fact_lepore?currentPage=all, accessed Sept. 13, 2014.

192 *Craft guilds almost all male:* Maryanne Kowaleski and Judith M. Bennett, "Crafts, Gilds, and Women in the Middle Ages: Fifty Years After Marian K. Dale," *Signs: Journal of Women in Culture and Society* 14, no. 21 (1989): 474–88.

192 *"white married women's labor force participation":* Raquel Fernández, "Cultural Change as Learning: The Evolution of Female Labor Force

Participation over a Century," *American Economic Review* 103, no. 1 (2013): 472.

193 *"In the first few decades of the twentieth century":* Dora L. Costa, "From Mill Town to Board Room: The Rise of Women's Paid Labor," *Journal of Economic Perspectives* 14, no. 4 (2000): 101.

193 *"As late as 1970":*Ibid., 101.

194 *"[They] began singing":* Irving Howe, *World of Our Fathers* (New York: Harcourt Brace, 1976), 299–300.

195 *"[A] young man helped a girl to the window sill":* Ibid., 305.

## Chapter 8: The Trouble with Men

200 *Womb envy:* Margaret Mead, *Male and Female* (New York: Morrow, 1949). Mead seems to have adopted the idea from psychoanalyst Karen Horney; for a history of the idea, see Miriam M. Johnson's *Strong Mothers, Weak Wives: The Search for Gender Equality* (Berkeley: University of California Press, 1988).

201 *Same baby, different responses:* J. A. Will, P. A. Self, and N. Datan, "Maternal Behavior and Perceived Sex of Infant," *American Journal of Orthopsychiatry* 46 (1976): 135–39.

201 *audio of a child's mischievous remarks:* Mary K. Rothbart and Eleanor Emmons Maccoby, "Parents' Differential Reactions to Sons and Daughters," *Journal of Personality and Social Psychology* 4 (1966): 237–43.

201 *PASTON Scale:* K. H. Karraker, D. A. Vogel, and M. A. Lake, "Parents' Gender-Stereotyped Perceptions of Newborns: The Eye of the Beholder Revisited," *Sex Roles* 33, no. 9–10 (1995): 687–701.

201 *A study of three-week old infants:* Howard Moss, "Sex, Age, and State as Determinants of Mother-Infant Interaction," *Merrill-Palmer Quarterly* 13 (1967): 19–36.

201 *Children label themselves:* The best overview of the similarities and differences between boys and girls in behavioral and psychological development is Eleanor Maccoby's *The Two Sexes: Growing Up Apart, Coming Together* (Cambridge, MA: Harvard University Press, 1998).

201 *Confirming older studies on other ethnic groups:* M. L. Halim, D. Ruble, C. Tamis-LeMonda, and P. E. Shrout, "Rigidity in Gender-Typed Behaviors in Early Childhood: A Longitudinal Study of Ethnic Minority Children," *Child Development* 84, no. 4 (2013): 1269–84.

202 *Same-sex playmate preference:* C. L. Martin, O. Kornienko, D. R. Schaefer, L. D. Hanish, R. A. Fabes, and P. Goble, "The Role of Sex of Peers and Gender-Typed Activities in Young Children's Peer Affiliative Networks: A Longitudinal Analysis of Selection and Influence," *Child Development* 84, no. 3 (2013): 921–37.

202 *"Jimmy's Baby Doll and Jenny's Truck":* C. Conry-Murray and E. Turiel, "Jimmy's Baby Doll and Jenny's Truck: Young Children's Reasoning About Gender Norms," *Child Development* 83, no. 1 (2012): 146–58.

202 *Same-sex groups as two cultures:* Maccoby, *The Two Sexes.*

202 *Over a third of babies watched television:* Common Sense Media, "Zero to Eight: Children's Media Use in America 2013," October 28, 2013, at http://www.commonsensemedia.org/research/zero-to-eight-childrens-media-use-in-america-2013, accessed Sept. 13, 2014.

203 *Child-training practices from ethnographic descriptions:* H. Barry, M. K. Bacon, and I. L. Child, "A Cross-Cultural Survey of Some Sex Differences in Socialization," *Journal of Abnormal and Social Psychology* 55 (1957): 327–32.

203 *Luo boys assigned girls' work:* Carol R. Ember, "Feminine Task Assignment and the Social Behavior of Boys," *Ethos* 1 (1973): 424–39.

203 *Big changes and new programs for fathering:* For details and references, see Melvin Konner, *The Evolution of Childhood: Relationships, Emotion, Mind* (Cambridge, MA: Harvard University Press, 2010), 475.

203 *Male monkeys become more nurturing:* G. Mitchell, "Paternalistic Behavior in Primates," *Psychological Bulletin* 71, no. 6 (June 1969): 399–417; and Gary Mitchell, William K. Redican, and Jody Gomber, "Lesson from a Primate: Males Can Raise Babies," *Psychology Today* 7, no. 11 (April 1974): 63–68.

204 *cultures with frequent combat separate fathers:* J. W. M. Whiting and B. B. Whiting, "Aloofness and Intimacy Between Husbands and Wives," *Ethos* 3 (1975): 183–207.

204 *Girls' and boys' cultures diverge by adolescence:* Alice Schlegel and Herbert Barry III, *Adolescence: An Anthropological Inquiry* (New York: Free Press, 1991).

205 *Children of gay and lesbian parents:* Charlotte J. Patterson, "Children of Lesbian and Gay Parents: Psychology, Law and Policy," *American Psychologist* 64, no. 8 (2009): 727–36. See also Konner, *Evolution of Childhood,* 329–34.

206 *Montagu's* Natural Superiority: Ashley Montagu, *The Natural Superiority of Women,* 5th ed. (Walnut Creek, CA: SAGE, 1999).

207 *Diana Nyad's swim from Cuba to Florida:* For a biography and interview, see *National Geographic,* which named her one of its 2014 "Adventurers of the Year," at http://adventure.nationalgeographic.com/adventure/adventurers-of-the-year/2014/diana-nyad/, accessed April 10, 2014. For her inspiring TED talk, see http://www.ted.com/talks/diana_nyad_never_ever_give_up.

208 *Nobel Prizes in science:* Montagu, *Natural Superiority,* 50–51.

208 *Today we can add:* See http://www.nobelprize.org/nobel_prizes/lists/women.html for women Nobel laureates since the prize began.

208 *"May it not be":* Montagu, 50.

209 *Common brain defects affect boys more:* S. Trent and W. Davies, "The Influence of Sex-Linked Genetic Mechanisms on Attention and Impulsivity," *Biological Psychology* 89, no. 1 (2012): 1–13; and A. M. Bao and D. F. Swaab, "Sexual Differentiation of the Human Brain: Relation to Gender Identity, Sexual Orientation and Neuropsychiatric Disorders," *Frontiers in Neuroendocrinology* 32, no. 2 (2011): 214–26.

209 *"It's pretty difficult to find any single factor":* Quoted by Constance Holden, "Sex and the Suffering Brain," *Science* 308 (2005): 1574. Insel was also referring to the preponderance of depression, anxiety, and eating disorders in women.

209 *Life-course-persistent conduct disorder:* R. F. Eme, "Sex Differences in Child-Onset, Life-Course-Persistent Conduct Disorder. A Review of Biological Influences," *Clinical Psychology Review* 27, no. 5 (2007): 607–27.

209 *Males "win" hands down:* Montagu, *Natural Superiority,* 133.

210 *Genetic defects with older parents due to older fathers:* E. Callaway, "Genetics: Fathers Bequeath More Mutations as They Age," *Nature* 488, no. 7412 (2012): 439; Augustine Kong and twenty other authors, "Rate of De Novo Mutations and the Importance of Father's Age to Disease Risk," *Nature* 488, no. 7412 (2012): 471–75.

210 *Differences in life expectancy:* J. C. Regan and L. Partridge, "Gender and Longevity: Why Do Men Die Earlier Than Women? Comparative and Experimental Evidence," *Best Practice & Research Clinical Endocrinology & Metabolism* 27, no. 4 (2013): 467–79. Men may be catching up in longevity in some countries, viewed from age sixty-five rather than from birth: M. Thorslund, J. W. Wastesson, N. Agahi,

M. Lagergren, and M. G. Parker, "The Rise and Fall of Women's Advantage: A Comparison of National Trends in Life Expectancy at Age Sixty-five Years," *European Journal of Ageing* 10, no. 4 (2013): 271–77.

210 *Career time committed to the "mommy track":* This is not intended to imply that men should not share equally in child care, only that in families where that does not happen (for example, where there is no father), women's lower mortality and greater longevity can offset a significant part of lost career time, and mothers should not be targets of discrimination.

211 *Women gain influence after children grow up:* J. K. Brown, "Cross-Cultural Perspectives on Middle-Aged Women," *Current Anthropology* 23 (1982): 143–56.

212 *Difference at seventeen and twenty-nine months:* Raymond H. Baillargeon and seven other authors, "Gender Differences in Physical Aggression: A Prospective Population-Based Survey of Children Before and After 2 Years of Age," *Developmental Psychology* 43, no. 1 (2007): 13–26.

212 *persists throughout growth:* For a broad overview in evolutionary context, see J. Archer, "Does Sexual Selection Explain Human Sex Differences in Aggression?" *Behavioral and Brain Sciences* 32, nos. 3–4 (2009): 249–66; discussion on pp. 266–311.

212 *Predation has little to do with aggression:* For details, see Melvin Konner, *The Tangled Wing: Biological Constraints on the Human Spirit*, rev. ed. (New York: Holt/Times Books, 2002), 184–85. The complete references can be found at http://www.melvinkonner.com/images/PDFs/tangledwingnotes.pdf, accessed Sept. 13, 2014.

213 *Female aggression limited and different:* A. Campbell, "The Evolutionary Psychology of Women's Aggression," *Philosophical Transactions of the Royal Society B: Biological Sciences* 368, no. 1631 (2013): 20130078.

213 *Golding's novel:* William Golding, *Lord of the Flies* (London: Faber & Faber, 1954).

214 *Robbers Cave:* Muzafer Sherif, O. J. Harvey, B. Jack White, William R. Hood, and Carolyn W. Sherif, *Intergroup Conflict and Cooperation: The Robbers Cave Experiment* (Norman, OK: Institute of Group Relations, 1961).

214 *Many experiments with adults:* Peter Robinson and Henri Tajfel, eds.,

*Social Groups and Identities: Developing the Legacy of Henri Tajfel,* International Series in Social Psychology (London: Butterworth-Heinemann, 1997).

215 *"male warrior hypothesis":* M. M. McDonald, C. D. Navarrete, and M. Van Vugt, "Evolution and the Psychology of Intergroup Conflict: The Male Warrior Hypothesis," *Philosophical Transactions of the Royal Society: Biological Sciences* 367, no. 1589 (2012): 670–79.

215 *More negative opinions when fertility risk was high:* C. D. Navarrete, D. M. T. Fessler, D. S. Fleischman, and J. Geyer, "Race Bias Tracks Conception Risk Across the Menstrual Cycle," *Psychological Science* 20, no. 6 (2009): 661–65.

215 *not just the black-white divide:* M. M. McDonald, B. D. Asher, N. L. Kerr, and C. D. Navarrete, "Fertility and Intergroup Bias in Racial and Minimal-Group Contexts: Evidence for Shared Architecture," *Psychological Science* 22, no. 7 (2011): 860–65.

216 *"social conditioning plays a considerable role":* Montagu, *Natural Superiority,* 144.

217 *Power "is the great aphrodisiac":* Hedrick Smith, "Foreign Policy: Kissinger at the Hub," *New York Times,* January 19, 1971, p. 12, downloaded April 11, 2014, at http://query.nytimes.com/mem/archive/pdf?res=F50D11F7345C107B93CBA8178AD85F458785F9. The article, including the quotation, was also placed in the Congressional Record, January 26, 1971: http://www.mocavo.com/Congressional-Record-Volume-117-Cong-92-Sess-1-Part-1/857762/334, accessed Sept. 13, 2014.

217 *"What motivates people to succeed?":* Maneet Ahuja, *The Alpha Masters: Unlocking the Genius of the World's Top Hedge Funds* (New York: Wiley, 2012), 116–17.

217 *"I was just thinking, you know":* Question and answer transcribed by the author; video available at http://video.cnbc.com/gallery/?video=3000092902, accessed April 11, 2014.

218 *"We don't really like each other in person":* Kate Taylor, "Sex on Campus: She Can Play That Game, Too," *New York Times,* July 14, 2013, Sunday Styles section, p. 1; published online July 12, 2013, at http://www.nytimes.com/2013/07/14/fashion/sex-on-campus-she-can-play-that-game-too.html?pagewanted=all, accessed Sept. 13, 2014.

218 *"I can't just lose my V-card":* Ibid., 7.

218 *"There's girls dancing in the middle":* Ibid., 7.

219 *"'Get down on your knees'":* Ibid., 7.

219 *"The Morning After":* Anne Campbell, "The Morning After the Night Before: Affective Reactions to One-Night Stands Among Mated and Unmated Women and Men," *Human Nature* 19, no. 2 (2008): 157–73.

219 *"The men had subsequently behaved disrespectfully":* Ibid., 162.

220 *"I have a very poor self image":* Ibid., 166.

220 *Men's comments:* Ibid., 165.

220 *"Thought it would be one of life's experiences":* Ibid., 166.

220 *Hookups at Syracuse University:* J. M. Townsend and T. H. Wasserman, "Sexual Hookups Among College Students: Sex Differences in Emotional Reactions," *Archives of Sexual Behavior* 40, no. 6 (2011): 1173–81.

221 *Kinsey Institute review of studies of hookup culture:* Justin R. Garcia, Chris Reiber, Sean G. Massey, and Ann M. Merriwether, "Sexual Hookup Culture: A Review," *Review of General Psychology* 16, no. 2 (2012): 161–76.

221 *"Bare Market":* Jeremy E. Uecker and Mark D. Regnerus, "Bare Market: Campus Sex Ratios, Romantic Relationships, and Sexual Behavior," *Sociological Quarterly* 51 (2010): 408–35.

222 *Similar effect in high schools:* W. D. Manning, M. A. Longmore, and P. C. Giordano, "Adolescents' Involvement in Non-Romantic Sexual Activity," *Social Science Research* 34, no. 2 (2005): 384–407.

222 *"Women possess something very important":* Shostak, *Nisa,* 257.

222 *"Among all peoples it is primarily men":* Donald Symons, *The Evolution of Human Sexuality* (New York: Oxford University Press, 1979), 253.

223 *"Men . . . have more frequent and more intense sexual desires":* Roy F. Baumeister, Kathleen R. Catanese, and Kathleen D. Vohs, "Is There a Gender Difference in Strength of Sex Drive? Theoretical Views, Conceptual Distinctions, and a Review of Relevant Evidence," *Personality and Social Psychology Review* 5, no. 3 (2001): 242.

223 *An Australian survey:* Ibid., 246.

223 *Classic study of sexual offers:* Russell D. Clark and Elaine Hatfield, "Gender Differences in Receptivity to Sexual Offers," *Journal of Psychology and Human Sexuality* 2, no. 1 (1989): 39–55. For an entertaining and unapologetic retrospective, see their "Love in the Afternoon," *Psychological Inquiry* 14, nos. 3–4 (2003): 227–31.

224 *2011 "replication":* T. D. Conley, A. C. Moors, J. L. Matsick, A.

Ziegler, and B. A. Valentine, "Women, Men, and the Bedroom: Methodological and Conceptual Insights That Narrow, Reframe, and Eliminate Gender Differences in Sexuality," *Current Directions in Psychological Science* 20, no. 5 (2011): 296–300.

224 *Petersen and Hyde confirmed substantial differences:* J. L. Petersen and J. S. Hyde, "Gender Differences in Sexual Attitudes and Behaviors: A Review of Meta-Analytic Results and Large Datasets," *Journal of Sex Research* 48, nos. 2–3 (2011): 149–65.

224 *Sex differences in fantasy lives:* Bruce J. Ellis and Donald Symons, "Sex Differences in Sexual Fantasy: An Evolutionary Psychological Approach," *Journal of Sex Research* 27, no. 4 (1990): 527–55.

225 *Lesbian versus gay male relationships:* Phillip Blumstein and Pepper Schwartz, *American Couples: Money, Work, Sex* (New York: William Morrow, 1983).

225 *STD risk in gay, heterosexual, and lesbian relationships:* B. G. Everett, "Sexual Orientation Disparities in Sexually Transmitted Infections: Examining the Intersection Between Sexual Identity and Sexual Behavior," *Archives of Sexual Behavior* 42, no. 2 (2013): 225–36; F. Xu, M. R. Sternberg, and L. E. Markowitz, "Men Who Have Sex with Men in the United States: Demographic and Behavioral Characteristics and Prevalence of HIV and HSV-2 Infection; Results from National Health and Nutrition Examination Survey 2001–2006," *Sexually Transmitted Diseases* 37, no. 6 (2010): 399–405; F. Xu, M. R. Sternberg, and L. E. Markowitz, "Women Who Have Sex with Women in the United States: Prevalence, Sexual Behavior and Prevalence of Herpes Simplex Virus Type 2 Infection; Results from National Health and Nutrition Examination Survey 2001–2006," *Sexually Transmitted Diseases* 37, no. 7 (2010): 407–13.

225 *Who pays cash for sex?:* See discussion and references under "Sexual Economics" below.

226 *"Romance and pornography":* Catherine Salmon, "The Pop Culture of Sex: An Evolutionary Window on the Worlds of Pornography and Romance," *Review of General Psychology* 16, no. 2 (2012): 152.

226 *"Pornography is a male fantasy world":* Ibid., 158.

226 Fifty Shades *novel:* E. L. James, *Fifty Shades of Grey* (New York: Vintage, 2012). They are reconciled in the sequel, but on somewhat different terms.

226 *"Sexual Economics":* Roy F. Baumeister and Kathleen D. Vohs,

"Sexual Economics: Sex as Female Resource for Social Exchange in Heterosexual Interactions," *Personality & Social Psychology Review* 8, no. 4 (2004): 339–63.

227 *Charlie Sheen and the Hollywood Madam:* Shawn Hubler, "Actor Says He Got Call Girls from Fleiss on at Least 27 Occasions: Trial: Jury Views Videotape of Testimony by Fidgeting Charlie Sheen; Checks Totaled More Than $50,000," *Los Angeles Times,* July 21, 1995, http://articles.latimes.com/1995-07-21/local/me-26278_1_dr-paul-fleiss-heidi-fleiss-tax-evasion-and-money, accessed Sept. 13, 2014.

227 *Strip clubs mainly women performing for men:* For details and references, see the Wikipedia entry "strip club," http://en.wikipedia.org/wiki/Strip_club, accessed Sept. 13, 2014.

227 *Coercive sex is overwhelmingly male:* Lawrence A. Greenfeld, *Sex Offenses and Offenders: An Analysis of Data on Rape and Sexual Assault,* U.S. Department of Justice, Bureau of Justice Statistics, February 1997, downloaded Sept. 13, 2014, at http://www.bjs.gov/content/pub/pdf/SOO.PDF.

228 *Psychologist Melissa Hines:* M. Hines, "Prenatal Endocrine Influences on Sexual Orientation and on Sexually Differentiated Childhood Behavior," *Frontiers in Neuroendocrinology* 32, no. 2 (2011): 170–82.

228 *Neurobiologist Margaret McCarthy:* M. M. McCarthy, "A Lumpers Versus Splitters Approach to Sexual Differentiation of the Brain," *Frontiers in Neuroendocrinology* 32, no. 2 (2011): 114–23.

228 *exposure to high levels of prenatal androgens:* S. A. Berenbaum and A. M. Beltz, "Sexual Differentiation of Human Behavior: Effects of Prenatal and Pubertal Organizational Hormones," *Frontiers in Neuroendocrinology* 32, no. 2 (2011): 183–200.

228 *"programmed into our brain":* A. M. Bao and D. F. Swaab, "Sexual Differentiation of the Human Brain: Relation to Gender Identity, Sexual Orientation and Neuropsychiatric Disorders," *Frontiers in Neuroendocrinology* 32, no. 2 (2011): 214.

229 *"little direct evidence supports this notion":* S. LeVay, "From Mice to Men: Biological Factors in the Development of Sexuality," *Frontiers in Neuroendocrinology* 32, no. 2 (2011): 112.

229 *differences beyond the hypothalamus:* J. Sacher, J. Neumann, H. Okon-Singer, S. Gotowiec, and A. Villringer, "Sexual Dimorphism in the Human Brain: Evidence from Neuroimaging," *Magnetic Resonance Imaging* 31, no. 3 (2013): 366–75.

229 *126 brain-imaging studies:* A. N. Ruigrok, G. Salimi-Khorshidi, M. C. Lai, S. Baron-Cohen, M. V. Lombardo, R. J. Tait, and J. Suckling, "A Meta-Analysis of Sex Differences in Human Brain Structure," *Neuroscience & Biobehavioral Reviews* 39, (2014): 34–50.

229 *changes in brain activity over the menstrual cycle:* J. Sacher, H. Okon-Singer, and A. Villringer, "Evidence from Neuroimaging for the Role of the Menstrual Cycle in the Interplay of Emotion and Cognition," *Frontiers in Human Neuroscience* 7 (2013): 374.

229 *Male amygdala larger:* Ruigrok et al., "Meta-Analysis of Sex Differences"; J. E. Bramen, J. A. Hranilovich, R. E. Dahl, E. E. Forbes, J. Chen, A. W. Toga, I. D. Dinov, C. M. Worthman, and E. R. Sowell, "Puberty Influences Medial Temporal Lobe and Cortical Gray Matter Maturation Differently in Boys Than Girls Matched for Sexual Maturity," *Cerebral Cortex* 21, no. 3 (2011): 636–46; and X. N. Zuo and ten other authors, "Growing Together and Growing Apart: Regional and Sex Differences in the Lifespan Developmental Trajectories of Functional Homotopy," *Journal of Neuroscience* 30, no. 45 (2010): 15034–43.

230 *Landmark menstrual cycle study:* Alice Rossi and Peter Rossi, "Body Time and Social Time: Mood Patterns by Menstrual Cycle Phase and Day of the Week," *Social Science Research* 6 (1977): 273–308.

## Chapter 9: Developing Daughters

231 *Rosin's provokingly titled book:* Hanna Rosin, *The End of Men: And the Rise of Women* (New York: Riverhead, 2012).

232 *Rosin, they said, grossly exaggerated:* See, for example, Jennifer Homans, "A Woman's Place," *New York Times Book Review,* September 13, 2012, and Stephanie Coontz, "The Myth of Male Decline," *New York Times,* September 29, 2012, http://www.nytimes.com/2012/09/30/opinion/sunday/the-myth-of-male-decline.html?pagewanted=all, accessed Sept. 13, 2014.

232 *CEOs of Fortune 500 companies:* See current data, accessed Sept. 13, 2014, at http://www.catalyst.org/knowledge/women-ceos-fortune-1000. The statistics are dynamic; the last time I had checked, in December 2013, the percentage of women among CEOs of Fortune 500 companies was 4.2, not 4.8.

234 *Georgia's WIN List:* http://gawinlist.com, accessed Sept. 13, 2014.

235 *EMILY's List:* http://www.emilyslist.org, accessed Sept. 13, 2014.

235 *Obama on the fiftieth anniversary of King's speech:* For video and transcript, see http://www.motherjones.com/mojo/2013/08/video-and-transcript-president-obama-speech-50th-anniversary-mlk-i-have-dream, accessed Sept. 13, 2014.

235 *Hillary Clinton concession speech:* For a transcript and a link to a video, see http://www.theguardian.com/commentisfree/2008/jun/07/hillary clinton.uselections20081, accessed Sept. 14, 2014.

236 *"Madam President":* http://www.emilyslist.org/news/entry/madam-president-iowa-town-hall, accessed Sept. 14, 2014. For the "deep bench" of future women candidates for president, see http://mpotus .tumblr.com/.

236 *Women heads of state:* For a brief history of women as national leaders, see http://content.time.com/time/photogallery/ 0,29307,2005290,00.html, accessed Sept. 14, 2014.

237 *Norwegian prime minister and cabinet:* http://www.regjeringen .no/en/the-government/solberg/members-of-the-government-2 .html?id=543170, accessed Sept. 14, 2014.

237 *UN Women's* Transformative Stand-Alone Goal: For a summary and links to the complete document, see http://www.unwomen .org/en/news/stories/2013/6/un-women-launches-global-call-for-a-transformative-agenda-to-make-gender-equality-a-reality/, accessed Sept. 14, 2014.

238 *"These ideas are not new": A Transformative Stand-Alone Goal,* p. 34, downloaded Sept. 14, 2014, from http://www.ipu.org/splz-e/ unga13/women.pdf.

238 *the best way to spend a development aid dollar:* For the basic argument, see Nancy Gibbs, "To Fight Poverty, Invest in Girls," *Time,* February 14, 2011, http://content.time.com/time/magazine/ article/0,9171,2046045,00.html, accessed April 13, 2014. Melinda French Gates, cofounder of the Gates Foundation, gives up-to-date references in "Putting Women and Girls at the Center of Development," *Science* 345, no. 6202 (2014): 1273–75.

239 *Sex ratios by age and country:* U.S. Central Intelligence Agency, *The World FactBook,* Field Listing: "Sex Ratio," accessed Sept. 14, 2014, at https://www.cia.gov/library/publications/the-world-factbook/fiel ds/2018.html.

239 *Something is wrong here:* T. Hesketh and Z. W. Xing, "Abnormal

Sex Ratios in Human Populations: Causes and Consequences," *Proceedings of the National Academy of Sciences of the United States of America* 103, no. 36 (2006): 13271–75.

240 *Missing women:* Amartya Sen, "More Than 100 Million Women Are Missing," *New York Review of Books,* December 20, 1990, accessed Sept. 14, 2014, at http://www.nybooks.com/articles/archives/1990/dec/20/more-than-100-million-women-are-missing/; and A. Sen, "Missing Women—Revisited: Reduction in Female Mortality Has Been Counterbalanced by Sex Selective Abortions," *British Medical Journal* 327, no. 7427 (2003): 1297–98.

240 *Skewed sex ratios in evolution:* R.A. Fisher, *The Genetical Theory of Natural Selection* (New York: Dover, 1958); and W. D. Hamilton, "Extraordinary Sex Ratios," *Science* 156, no. 3774 (1967): 477–88.

240 *Discouraging sex determination:* Kate Gilles and Charlotte Feldman-Jacobs, *When Technology and Tradition Collide: From Gender Bias to Sex Selection*, Population Reference Bureau, Policy Brief, September 2012, downloaded Sept. 14, 2014, from http://www.prb.org/Publications/Reports/2012/sex-selection.aspx.

240 *Banking money on a girl's birth:* Ibid., 4.

240 Atmajaa *TV series:* Ibid., 5.

240 *South Korean son preference declines:* Ibid., 4.

241 *son preference in South Korea "is over":* Monica Das Gupta quoted by Rosin, *The End of Men,* 235.

241 *Plummeting Bangladesh births:* Ruth Levine, Molly Kinder, and the "What Works?" Working Group, *Case Studies in Global Health: Millions Saved* (Sudbury, MA: Jones and Bartlett, 2004), Case 13: "Reducing Fertility in Bangladesh," pp. 97–104. PDF available for download at the Center for Global Development website, http://www.cgdev.org/page/case-13-reducing-fertility-bangladesh; 2012 data from http://data.worldbank.org/indicator/SP.DYN.TFRT.IN; both accessed Sept. 14, 2014.

241 *Education and media made a difference:* A. Goni and M. Rahman, "The Impact of Education and Media on Contraceptive Use in Bangladesh: A Multivariate Analysis," *International Journal of Nursing Practice* 18, no. 6 (2012): 565–73.

241 *Soap opera about Laila:* Levine, Kinder, et al., *Case Studies,* 5.

241 *Worldwide maternal mortality:* World Health Organization summary, accessed Sept. 14, 2014, at http://www.who.int/mediacentre/factsheets/fs348/en/.

241 *Sri Lanka success story:* Levine, Kinder, et al., *Case Studies,* Case 6: "Saving Mothers' Lives in Sri Lanka."

242 *"The things that kill mothers":* Interview with Lynn Sibley, "Maternal and Newborn Health in Ethiopia Partnership." Video available at http://www.international.emory.edu/awards/creekmore/sibley_video.html, accessed Sept. 14, 2014. The quote begins at minute 1:25. For background see http://www.international.emory.edu/awards/creekmore/previous/lynn_sibley.html and L. Sibley, S. T. Buffington, L. Tedessa Sr., and K. McNatt, "Home-Based Life Saving Skills in Ethiopia: An Update on the Second Phase of Field Testing," *Journal of Midwifery & Women's Health* 51, no. 4 (2006): 284–91.

243 *"Basically, no woman should die giving birth":* Sibley interview video, minute 3:05.

243 *Thailand HIV success story:* Levine, Kinder, et al., *Case Studies,* Case 2: "Preventing HIV/AIDS and Sexually Transmitted Infections in Thailand."

244 *pop-up models of penises:* See the PBS/BBC television series *Medicine at the Crossroads* (Thirteen/WNET, New York, and BBC, England, in collaboration with Televisión Española, SA, and the Australian Broadcasting Corporation and WETA, Washington, DC, 1993), episode 7 "Pandemic," and my book that accompanied the series, *Medicine at the Crossroads: The Crisis in Health Care* (New York: Pantheon, 1993), chapter 8.

244 *"The government has fallen asleep at the wheel":* Mechai Viravaidya quoted in "Thailand Faces New HIV/Aids Crisis," *Bangkok Post,* December 21, 2012, http://www.bangkokpost.com/breaking-news/327319/mechai-warns-of-new-hiv-aids-crisis, accessed Sept. 14, 2014.

244 *something else was going on in Thailand:* Elizabeth Pisani, *The Wisdom of Whores: Bureaucrats, Brothels, and the Business of Aids* (New York: W. W. Norton, 2009).

245 *a 2012 survey in Uganda:* Josh Kron, "In Uganda, an AIDS Success Story Comes Undone," *New York Times,* August 2, 2012, accessed Sept. 14, 2014, at http://www.nytimes.com/2012/08/03/world/africa/in-uganda-an-aids-success-story-comes-undone.html?_r=0.

245 *"dry sex":* T. Hull, ten other authors, and the WHO GSVP Study Group, "Prevalence, Motivations, and Adverse Effects of Vaginal Practices in Africa and Asia: Findings from a Multicountry Household

Survey," *Journal of Women's Health* 20, no. 7 (2011): 1097–1109. For Indonesia, see Trish Anderson's "Some Like It Dry," *Jakarta Post,* February 26, 2008, accessed Sept. 14, 2014, at http://www.thejakartapost.com/news/2008/02/26/some-it-dry.html.

245 *"Dry sex aids are the inverse of lube":* Coca Colo's *Femonomics* blog, entry for January 26, 2010, "What Is Dry Sex, and Why Do You Need to Know About It?" http://femonomics.blogspot.com/2010/01/what-is-dry-sex-and-why-do-you-need-to.html, accessed Sept. 14, 2014.

245 *Spread of HIV/AIDS in Africa:* E. Deuchert and S. Brody, "Lack of Autodisable Syringe Use and Health Care Indicators Are Associated with High HIV Prevalence: An International Ecologic Analysis," *Annals of Epidemiology* 17, no. 3 (2007): 199–207; and Devon D. Brewer, David Gisselquist, Stuart Brody, and John J. Potterat, "Investigating Iatrogenic HIV Transmission in Ugandan Children," *JAIDS—Journal of Acquired Immune Deficiency Syndromes* 45, no. 2 (2007): 253–54.

246 *"A second approach is male circumcision":* Bill Gates, "Annual Letter 2012," Bill & Melinda Gates Foundation, http://www.gatesfoundation.org/Who-We-Are/Resources-and-Media/Annual-Letters-List/Annual-Letter-2012, accessed Sept. 14, 2014.

247 *Delhi rape:* "Nirbhaya Gang-Rape Case: Delhi HC Upholds Death Penalty Awarded to 4 Convicts," *Times of India,* March 13, 2014, accessed Sept. 14, 2014, http://timesofindia.indiatimes.com/city/delhi/Nirbhaya-gang-rape-case-Delhi-HC-upholds-death-penalty-awarded-to-4-convicts/articleshow/31938726.cms.

247 *Mumbai rape:* Rebecca Samervel, "Mumbai Shakti Mills Rape Cases: Death Penalty for 3 Repeat Offenders," *Times of India,* April 4, 2014, accessed at http://timesofindia.indiatimes.com/city/mumbai/Mumbai-Shakti-Mills-rape-cases-Death-penalty-for-3-repeat-offenders/articleshow/33238680.cms, Sept. 14, 2014.

248 *Rape and beating of Swiss couple:* "Five Men Admit to Raping Swiss Tourist, Say Indian Police," Associated Press/Guardian, March 18, 2013, accessed at http://www.theguardian.com/world/2013/mar/18/admit-raping-swiss-tourist-india, Sept. 14, 2014; and "India Jails Six over Swiss Gang Rape in Madhya Pradesh," *BBC News India,* July 20, 2013, http://www.bbc.com/news/world-asia-india-23390078.

248 *WHO ten-country study:* Claudia García-Moreno, Henrica A. F. M. Jansen, Charlotte Watts, Mary Caroll Ellsberg, Lori Heise, and the

Gender, Women and Health Network, "WHO Multi-Country Study on Women's Health and Domestic Violence Against Women: Initial Results on Prevalence, Health Outcomes and Women's Responses," World Health Organization, 2012, accessed Sept. 14, 2014, at http://www.who.int/gender/violence/who_multicountry_study/en/. See also "Violence Against Women," WHO Media Centre Fact Sheet No. 239, updated October 2013, at http://www.who.int/mediacentre/factsheets/fs239/en/, accessed Sept. 14, 2014.

248 *London School of Tropical Hygiene further analysis:* K. M. Devries and thirteen other authors, "The Global Prevalence of Intimate Partner Violence Against Women," *Science* 340, no. 6140 (2013): 1527–28.

249 *New national law, but marital rape not a crime:* See the Wikipedia entry "The Criminal Law (Amendment) Act, 2013," accessed Sept. 14, 2014, at http://en.wikipedia.org/wiki/Criminal_Law_(Amendment)_Act,_2013.

249 *"Boys make mistakes":* "Mulayam Singh Yadav on Rape: Boys Make Mistakes, Shouldn't Hang," *India Today,* April 10, 2014, accessed Sept. 14, 2014, at http://indiatoday.intoday.in/story/mulayam-singh-yadav-on-rape-moradabad-anti-rape-laws-samajwadi-party-lok-sabha-election-2014/1/354991.html.

249 *"Both should be hanged":* "Women Having Sex Should Be Hanged: Abu Azmi Shocker on Yadav's Rape Remark," *India Today,* April 11, 2014, accessed Sept. 14, 2014, at http://indiatoday.intoday.in/story/abu-azmi-shocker-on-yadavs-rape-remark-women-having-sex-should-be-hanged/1/355087.html.

249 *Two girls raped and hanged:* Harmeet Shah Singh and Faith Karimi, "3 suspects Confess in India Gang Rape; Community Outraged," *CNNWorld,* June 4, 2014, accessed Sept. 14, 2014, at http://www.cnn.com/2014/06/01/world/asia/india-gang-rape/.

249 *"What's it to you?"* and *"Sometimes it's right":* Rob Verger, "Indian Official Says Rape Is 'Sometimes' Right," *Newsweek,* June 5, 2014, accessed Sept. 14, 2014, at http://www.newsweek.com/indian-official-says-rape-sometimes-right-253617.

250 *National comparisons of recorded rapes:* NationMaster website, http://www.nationmaster.com/country-info/stats/Crime/Violent-crime/Rapes, accessed Sept. 14, 2014, and Tom Wright, "Are Women Safer in India or the U.S.?" *Wall Street Journal,* January 2, 2013.

250 *Missoula, Montana, prosecutors:* Jack Healy, "Accusation in Montana of

Treating Rape Lightly Stirs Unlikely Public Fight," *New York Times*, April 12, 2014, http://www.nytimes.com/2014/04/13/us/accusation-in-montana-of-treating-rape-lightly-stirs-unlikely-public-fight.html?_r=0, accessed Sept. 14, 2014. The U.S. Department of Justice report, dated February 14, 2014, was also accessed on Sept.14, 2014, at http://www.justice.gov/crt/about/spl/documents/missoula_ltr_2-14-14.pdf.

250 *Steubenville rape, apologetics, victim blaming, and cover-up:* Juliet Macur and Nate Schweber, "Rape Case Unfolds on Web and Splits City," *New York Times*, December 16, 2012, accessed Sept. 14, 2014, at http://www.nytimes.com/2012/12/17/sports/high-school-football-rape-case-unfolds-online-and-divides-steubenville-ohio.html?pagewanted=all; and Laurie Penny, "Steubenville: This Is Rape Culture's Abu Ghraib Moment," *NewStatesman*, March 19, 2013, accessed Sept. 14, 2014, at http://www.newstatesman.com/laurie-penny/2013/03/steubenville-rape-cultures-abu-ghraib-moment. Four adults, including the school district superintendent, were indicted for negligence or covering up, but the coach who threatened reporters had his contract renewed and had faced no formal consequences as of June 7, 2014. The rapists, who were sixteen and seventeen at the time of the crime, were delinquent of rape (the juvenile equivalent of guilty) and sentenced to two years and one year, respectively. The first was still serving as of June 7, 2014; the second was released in January 2014 after serving ten months. Both will be listed as Tier II sex offenders.

251 *"The global pandemic of violence against women":* Ban Ki-moon quoted in the One Billion Rising press release of February 21, 2014, accessed Sept. 14, 2014, at http://www.onebillionrising.org/11825/press-release/.

251 *Female genital cutting:* For an overview, see "Female Genital Mutilation/Cutting: A Statistical Overview and Exploration of the Dynamics of Change," UNICEF, Statistics and Monitoring Section, Division of Policy and Strategy, July 2013, http://www.unicef.org/media/files/FGCM_Lo_res.pdf, downloaded Sept. 14, 2014. For academic and medical evidence, see H. L. Sipsma, P. G. Chen, A. Ofori-Atta, U. O. Ilozumba, K. Karfo, and E. H. Bradley, "Female Genital Cutting: Current Practices and Beliefs in Western Africa," *Bulletin of the World Health Organization* 90, no. 2 (2012): 120–27; T. H. Anis, S. A. Gheit, H. H. Awad, and H. S. Saied, "Effects of Female Genital Cutting on the Sexual Function of Egyptian

Women. A Cross-Sectional Study," *Journal of Sexual Medicine* 9, no. 10 (2012): 2682–92; R. Chibber, E. El-Saleh, and J. El Harmi, "Female Circumcision: Obstetrical and Psychological Sequelae Continues Unabated in the 21st Century," *Journal of Maternal-Fetal & Neonatal Medicine* 24, no. 6 (2011): 833–36.

251 *Orchid Project:* See http://orchidproject.org, accessed Sept. 14, 2014.

252 *"vacation cutting":* Julie Turkewitz, "A Fight as U.S. Girls Face Genital Cutting Abroad," *New York Times,* June 11, 2014, A1; also accessed online on Sept. 14, 2014, at http://www.nytimes.com/2014/06/11/us/a-fight-as-us-girls-face-genital-cutting-abroad.html?hp&_r=0.

252 *The grand mufti's fatwa:* He is quoted in "Fatwas Against FGM," at http://stopfgmmiddleeast.wordpress.com/fatwas-against-fgm/, accessed Sept. 14, 2014.

252 *The former president's book:* Jimmy Carter, *A Call to Action: Women, Religion, Violence, and Power* (New York: Simon and Schuster, 2014).

252 *Molly Melching and Tostan:* http://www.tostan.org. See also Aimee Molloy's *However Long the Night: Molly Melching's Journey to Help Millions of African Women and Girls Triumph* (New York: HarperOne, 2013).

252 *"understandable that people are outraged":* Melinda Gates, "Five Questions for Molly Melching," website of the Bill & Melinda Gates Foundation, June 25, 2013, http://www.impatientoptimists.org/Posts/2013/06/Melinda-Gates-5-Questions-for-Tostans-Molly-Melching, accessed Sept. 14, 2014.

253 *"Women's Empowerment":* The report is available at the SlideShare website, http://www.slideshare.net/Indian-CAG/young4leaders, accessed Sept. 14, 2014. Most of the information in this section, including the quotes, is based on this report, which I accessed on the CARE website in September 2013. Although it is available at this writing on SlideShare, I have found other sources for the remaining facts and quotes, as indicated below.

253 *"A key breakthrough in CARE's evolving understanding":* See "Women's Empowerment," 3, box.

253 *CARE's rebranding:* V. Kasturi Rangan and Katharine Lee, "Repositioning CARE USA," Harvard Business School Case 509–005, August 2008 (revised July 2009), at http://www.hbs.edu/faculty/Pages/item.aspx?num=36294, accessed Sept. 14, 2014; "CARE 'I Am

Powerful' Campaign," special issue of *Advertising & Society Review* 9, no. 1, 2008, accessed Sept. 14, 2014, at http://muse.jhu.edu/journals/advertising_and_society_review/toc/asr9.1.html.

254 *Educating girls:* See Gibbs, "To Fight Poverty, Invest in Girls"; Gates, "Putting Women and Girls at the Center of Development," cited above; and the Girl Effect website, www.girleffect.org, accessed Sept. 14, 2014.

254 *Benin girls:* "Women's Empowerment," 6–7; Care Internationale au Bénin, "Rapport de la Revue mi-parcours: Projet de Promotion de la Participation des Communautés pour une Education de Base pour Tous," accessed Sept. 14, 2014, at http://www.careevaluations.org/Evaluations/PROBASE%20Mid%20term%20Evaluation%20Dec%202003%20-%20Benin.doc.

254 *"smart, engaging, and bold girls":* "Women's Empowerment," 7.

255 *CARE "Microfinance-Plus":* For the CARE approach to microfinance, see http://www.careinternational.org.uk/what-we-do/microfinance, accessed Sept. 14, 2014. For a critique and evaluation of the standard microfinance approach, see D. McKenzie, "Impact Assessments in Finance and Private Sector Development: What Have We Learned and What Should We Learn?," *The World Bank Research Observer* 25, no. 2 (2009): 209–33.

255 *Women on the Move program:* W. J. Grant and H. Allen, "CARE's Mata Masa Dubara (Women on the Move) Program in Niger," *Journal of Microfinance* 4, no. 2 (2002).

256 *"Like you, we have"* and *"We realized we are also human":* quotations are from interviews described in Michael Drinkwater (CARE International), "'We Are Also Human': Identity and Power in Gender Relations," paper submitted to the conference "The Winners and Losers from Rights-Based Approaches to Development," University of Manchester, February 21–22, 2005, 4. See also "Women's Empowerment," 10.

256 *Statistics on human trafficking:* United Nations Office on Drugs and Crime, *Global Report on Trafficking in Persons 2012,* UNODC, Vienna; and U.S. Department of State, *Trafficking in Persons Report,* June 2013, Washington, D.C.

257 *Kidnapped Nigerian schoolgirls called slaves:* Adam Nossiter, "Nigerian Islamist Leader Threatens to Sell Kidnapped Girls," *New York Times,* May 5, 2014, accessed Sept. 14, 2014, at http://www

.nytimes.com/2014/05/06/world/africa/nigeria-kidnapped-girls.html.

257 *Girls' location approximately known*: Arwa Damon, Brent Swails, and Nick Thompson, "Where Are Nigeria's Missing Girls? On the Hunt for Boko Haram," *CNNWorld*, June 10, 2014, accessed on that date at http://www.cnn.com/2014/06/10/world/africa/boko-haram-hunt-arwa-damon/.

257 *Village attacked and girls dispersed:* Associated Press, "11 Parents of Nigeria's Kidnapped Girls Die from Attacks and Stress," July 22, 2014, accessed Sept. 14, 2014, at Fox News World, http://www.foxnews.com/world/2014/07/22/11-parents-nigeria-kidnapped-girls-die-from-attacks-and-stress-hometown-is/; and Aminu Abubakar, "Boko Haram engaged in talks over kidnapped girls," CNN World, September 9, 2012, accessed Sept. 23, 2014, at http://www.cnn.com/2014/09/20/world/africa/nigeria-boko-haram-kidnapped-girls/.

257 *"Malala":* Jon Boone, "Men Involved in Malala Yousafzai Shooting Arrested in Pakistan," *The Guardian*, Sept. 12, 2014, accessed Sept. 14, 2014, at http://www.theguardian.com/world/2014/sep/12/men-malala-yousafzai-shooting-arrested-pakistan.

257 *ISIS rapes and enslaves thousands:* Alessandria Masi, "ISIS Rapes, Tortures, Marries Captured Yazidi Women as Young as 13: State Department Report," *International Business Times*, Sept. 12, 2014, accessed Sept. 14, 2014, at http://www.ibtimes.com/isis-rapes-tortures-marries-captured-yazidi-women-young-13-state-department-report-1687394.

258 *The CNN Freedom Project:* http://thecnnfreedomproject.blogs.cnn.com, accessed Sept. 14, 2014.

258 *Free the Slaves:* http://www.freetheslaves.net, accessed Sept. 14, 2014.

258 *Free2Work:* http://www.free2work.org, accessed Sept. 14, 2014.

258 *Not for Sale:* http://www.notforsalecampaign.org, accessed Sept. 14, 2014.

258 *Coalition Against Trafficking in Women:* http://www.catwinternational.org, accessed Sept. 14, 2014.

258 *The International Princess Project:* http://intlprincess.org, accessed Sept. 14, 2014.

258 *CSEC in Atlanta:* Bob Herbert, "Young, Cold, and for Sale," *New York Times* op-ed page, October 19, 2006, accessed Sept. 14, 2014,

at http://www.nytimes.com/2006/10/19/opinion/19herbert.html?_r=0; and Alexandra Priebe and Cristen Suhr, *Hidden in Plain View: The Commercial Sexual Exploitation of Girls in Atlanta, Executive Summary,* a study of the Atlanta Women's Agenda, September 2005, downloaded Sept. 14, 2014 from www.childtrafficking.com/Docs/atlanta_women_05_girls_0109.pdf. For a video, see https://www.youtube.com/watch?v=tv6b5YDD92o, accessed Sept. 14, 2014.

259 *Project PREVENT:* http://schoolsafety.education.gsu.edu/research/project-p-r-e-v-e-n-t/, accessed Sept. 14, 2014, and A. Kruger, E. Harper, P. Harris, D. Sanders, K. Levin, and J. Meyers, "Sexualized and Dangerous Relationships: Listening to the Voices of Low-Income African American Girls Placed at Risk for Sexual Exploitation," *Western Journal of Emerging Medicine* 14, no. 4 (2013): 370–76.

259 *"It's a whole bunch of girls":* This and other quotes in this section are from Kruger et al., "Sexualized and Dangerous," 373.

260 *Georgia Care Connection estimates:* http://www.georgiacareconnection.com, accessed Sept. 14, 2014.

260 *Angela's House:* http://www.youth-spark.org/learn/our-programs/legacy-programs/angelas-house/, accessed Nov. 4, 2014.

260 *"Dear John" Campaign:* For background description and video, see http://shirley-franklin.com/?page_id=732, accessed Sept. 14, 2014.

260 *"CSEC 101":* See the webinar presented by Jennifer Felner and Amber McKeen of Children's Hospital of Atlanta, http://www.choa.org/childrens-hospital-services/child-protection-center/prevention-and-training/training-videos#prettyPhotoVideo70715[inline]/0/, accessed Nov. 4, 2014.

261 *National Human Trafficking Hotline:* http://www.polarisproject.org/what-we-do/national-human-trafficking-hotline/the-nhtrc/overview, accessed Sept. 14, 2014. The hotline is non-governmental despite the State of Georgia's posting requirement.

261 *"41 key legislative components":* http://sharedhope.org/what-we-do/bring-justice/reportcards/, accessed Sept. 14, 2014.

261 *CSEC milestones in Georgia:* http://bartoncenter.net/work/publications.html, accessed Sept. 14, 2014.

261 *California recognized Georgia:* http://children.georgia.gov/press-releases/2013-04-04/georgias-statewide-response-csec-recently-cited, accessed Sept. 14, 2014.

261 *"The deputy separated the young girl":* http://blogs.ajc.com/atlanta-forward/2013/01/17/human-trafficking/, accessed Sept. 14, 2014.

262 *Claude Steele and stereotype threat:* Claude M. Steele, "A Threat in the Air: How Stereotypes Shape Intellectual Identity and Performance," *American Psychologist* 52, no. 6 (1997): 613–29; and Claude M. Steele, Steven J. Spencer, and Joshua Aronson, "Contending with Group Image: The Psychology of Stereotype and Social Identity Threat," *Advances in Experimental Social Psychology* 34 (2003): 379–440. See also Steele's *Whistling Vivaldi: How Stereotypes Affect Us and What We Can Do* (New York: W. W. Norton, 2011).

262 *Stereotype threat in women:* Hannah-Hanh D. Nguyen and Ann Marie Ryan, "Does Stereotype Threat Affect Test Performance of Minorities and Women? A Meta-Analysis of Experimental Evidence," *Journal of Applied Psychology* 93, no. 6 (2008): 1314–34.

262 *Seventeen years of research on threat and gender equality:* Katherine Picho, Ariel Rodriguez, and Lauren Finnie, "Exploring the Moderating Role of Context on the Mathematics Performance of Females under Stereotype Threat: A Meta-Analysis," *Journal of Social Psychology* 153, no. 3 (2013): 299–333.

262 *Country differences in gender math gap:* Andrew M. Penner, "Gender Differences in Extreme Mathematical Achievement: An International Perspective on Biological and Social Factors," supplement, *American Journal of Sociology* 114 (2008): S138–S170.

262 *framing has a major impact:* Emily S. Shaffer, David M. Marx, and Radmila Prislin, "Mind the Gap: Framing of Women's Success and Representation in Stem Affects Women's Math Performance Under Threat," *Sex Roles* 68, nos. 7–8 (2012): 454–63.

262 *"Recent Study Shows Men and Women Nearly Equal":* Ibid., 461. Maryam Mirzakhani won a Fields Medal in 2014.

263 *"it is important to realize"* and *"more strides need to be made":* Ibid., 462.

263 *"Calling attention to women's success":* Ibid., 459.

263 *French middle school girls did worse:* Annique Smeding, Florence Dumas, Florence Loose, and Isabelle Regner, "Order of Administration of Math and Verbal Tests: An Ecological Intervention to Reduce Stereotype Threat on Girls' Math Performance," *Journal of Educational Psychology* 105, no. 3 (2013): 850–60.

263 *ninth graders and college students in Germany:* Melanie C. Steffens and Petra Jelenec, "Separating Implicit Gender Stereotypes Regarding

Math and Language: Implicit Ability Stereotypes Are Self-Serving for Boys and Men, but Not for Girls and Women," *Sex Roles* 64, nos. 5–6 (2011): 324–35.

263 *Single-sex and co-ed schools in Uganda:* Katherine Picho and Jason M. Stephens, "Culture, Context and Stereotype Threat: A Comparative Analysis of Young Ugandan Women in Coed and Single-Sex Schools," *Journal of Educational Research* 105, no. 1 (2012): 52–63.

264 *"Mathematics and Sciences Course Taking Among Arab Students":* Hanna Ayalon, "Mathematics and Sciences Course Taking Among Arab Students in Israel: A Case of Unexpected Gender Equality," *Educational Evaluation and Policy Analysis* 24, no. 1 (2002): 63–80.

264 *Israeli Arabs not equal citizens:* See, for example, http://nif.org/media-center/nif-news/1165-the-current-protests-are-a-qdream-come-trueq, at the website of the New Israel Fund, accessed Sept. 14, 2014. My own views on Israel and its conflicts, internal and external, are complex and cannot be captured here. If interested, please visit www.jewsan dothers.com. I strongly recommend *My Promised Land: The Triumph and Tragedy of Israel,* by Ari Shavit (New York: Spiegel & Grau, 2013); and *Once Upon a Country: A Palestinian Life,* by Sari Nusseibeh with Anthony David (New York: Farrar, Straus & Giroux, 2007).

264 *Except for one group of Jewish girls:* Yariv Feniger, "The Gender Gap in Advanced Math and Science Course Taking: Does Same-Sex Education Make a Difference?" *Sex Roles* 65, nos. 9–10 (2010): 670–79.

264 *Women at the top institutes of technology:* Amy Sue Bix, *Girls Coming to Tech! A History of American Engineering Education for Women* (Cambridge, MA: MIT Press, 2013).

265 *"Before 1952, Georgia Tech had no undergraduate women":* From Amy Bix, "'Women Are NOT for Engineering': How Caltech, Georgia Tech, and Other Schools Reluctantly Began Accepting Female Students into Engineering Programs," http://www.slate .com/articles/technology/future_tense/2014/03/women_in_ stem_how_georgia_tech_caltech_and_others_opened_up_to_ female_engineers.html, accessed Sept. 14, 2014.

265 *An Hour of Code:* See "4 Ways to Recruit Girls to Try Computer Science," at https://code.org/girls, accessed Sept. 14, 2014.

265 *Larry Summers on women in science:* Lois Romano, "Embattled Harvard President to Resign," *Washington Post,* February 22, 2006,

accessed Sept. 14, 2014, at http://www.washingtonpost.com/wp-dyn/content/article/2006/02/21/AR2006022101842.html.

265 *Janet Yellen beats rival:* Michael Riley and Jeff Kearns, "Summers Backs Yellen's Debut in Fed Chair Role He Wanted," *Bloomberg News,* February 16, 2014, at http://www.bloomberg.com/news/2014-02-15/yellen-says-recovery-in-labor-market-far-from-complete.html, accessed Sept. 14, 2014.

265 *priming women with a sense of their own power:* Katie J. Van Loo and Robert J. Rydell, "On the Experience of Feeling Powerful: Perceived Power Moderates the Effect of Stereotype Threat on Women's Math Performance," *Personality and Social Psychology Bulletin* 39, no. 3 (2013): 387–400.

## Chapter 10: Billions Rising

267 *The antics of Secret Service men:* CNN Wire Staff, "Secret Service Agents Relieved in Colombia Amid Prostitution Allegations," CNN World, April 14, 2012, http://www.cnn.com/2012/04/14/world/americas/colombia-summit-secret-service/, accessed Sept. 14, 2014. In the end the men got light punishments, and their actions do not seem to have been unusual: Shane Harris, "Secret Service Prostitution Scandal: One Year Later," *Washingtonian,* April 2013 issue, accessed Sept. 14, 2014, at http://www.washingtonian.com/articles/people/secret-service-prostitution-scandal-one-year-later/index.php.

267 *Former Senator's scandal and mistrial:* David Zucchino, "John Edwards Not Guilty on 1 Count, Mistrial Declared in Other 5," *Los Angeles Times,* May 31, 2012, accessed Sept. 14, 2014, at http://articles.latimes.com/2012/may/31/news/la-pn-john-edwards-trial-verdict-20120531. As of 2014 he had returned to his former career as a trial lawyer representing victims of alleged medical malpractice: Catalina Camia, "John Edwards Back in N.C. Court as a Trial Lawyer," *USA Today*, April 16, 2014, at http://onpolitics.usatoday.com/2014/04/16/john-edwards-trial-lawyer-medical-malpractice/, accessed Sept. 14, 2014.

268 *France had chosen a new leader:* The malefactor, Dominique Strauss-Kahn, is favored in some polls to replace the man at the top who replaced him, according to *France 24,* March 5, 2014, at http://www.france24.com/en/20140305-dsk-strauss-kahn-best-man-run-france-hollande-poll/, accessed Sept. 14, 2014. According

to the article, which summarizes the New York episode, Strauss-Kahn still faces charges in France for other alleged sex-related offenses. However, with a new film about him starring Gérard Depardieu and Jacqueline Bisset, despite its unflattering portrayal, he has made several public appearances and seems to continue to relish the limelight: Sylvie Kaufmann, "Why D.S.K. Won't Go Away," *New York Times,* May 24, 2014, accessed Sept. 14, 2014 at http://www.nytimes.com/2014/05/25/opinion/sunday/why-dsk-wont-go-away.html.

268 *San Diego mayor:* Josh Levs and David Simpson, "Ex-San Diego Mayor Sentenced to Home Confinement for Assaulting Women," *CNN Justice,* December 9, 2013, at http://www.cnn.com/2013/12/09/justice/ex-san-diego-mayor-bob-filner-sentencing/, accessed Sept. 14, 2014; and Laila Kearney, "Ex-San Diego Mayor Bob Filner released from house arrest," Reuters Edition: US, April 7, 2014, at http://www.reuters.com/article/2014/04/07/us-usa-california-filner-idUSBREA361SM20140407, accessed Sept. 14, 2014.

268 *The embattled mayor of Toronto:* Carol J. Williams, "Rob Ford quits Toronto mayor's race but plans to run for City Council," *Los Angeles Times,* Sept. 12, 2014, at http://www.latimes.com/world/mexico-americas/la-fg-rob-ford-withdraws-toronto-mayor-race-20140912-story.html, accessed Sept. 14, 2014.

269 *California governor fathers child by nanny:* CNN Wire Staff, "Schwarzenegger Admits Habit of Keeping Secrets, Including Multiple Affairs," *CNN U.S.,* October 2, 2012, at http://www.cnn.com/2012/09/30/us/schwarzenegger-interview/, accessed Sept. 14, 2014.

269 *New York governor hires prostitutes:* Michael M. Grynbaum, "Spitzer Resigns, Citing Personal Failings," *New York Times,* March 12, 2008, accessed Sept. 14, 2014, at http://www.nytimes.com/2008/03/12/nyregion/12cnd-resign.html?pagewanted=all; and Michael M. Grynbaum, "Spitzer and His Wife Say Their Marriage Is Over," *New York Times,* December 24, 2013, at http://www.nytimes.com/2013/12/25/nyregion/spitzer-and-his-wife-say-their-marriage-is-over.html?ref=eliotlspitzer&_r=0, accessed Sept. 14, 2014.

269 *New Jersey Governor admits harassment, resigns:* David Kocieniewski, "The Governor Resigns: The Decision; A Governor's Downfall, in 20 Wrenching Days," *New York Times,* August 15, 2004, accessed Sept. 14, 2014, at http://www.nytimes.com/2004/08/15/nyregion/the-

governor-resigns-the-decision-a-governor-s-downfall-in-20-wrench ing-days.html?ref=jamesemcgreevey. This story appears to have had a happy ending: Michelle Green, "James McGreevey, This Side of Redemption," *New York Times,* March 27, 2013, accessed Sept. 14, 2014, at http://www.nytimes.com/2013/03/28/fashion/this-side-of-redemption.html?ref=jamesemcgreevey.

269 *South Carolina governor out of touch:* Peter Hamby and Kristy Keck, "South Carolina Gov. Sanford Admits Extramarital Affair," *CNN Politics.com,* June 24, 2009, at http://www.cnn.com/2009/POLITICS/06/24/south.carolina.governor/, accessed Sept. 14, 2014.

269 *"Hiking the Appalachian Trail":* Kim Severson, "Looking Past Sex Scandal, South Carolina Returns Ex-Governor to Congress," *New York Times,* May 7, 2013, at http://www.nytimes.com/2013/05/08/us/south-carolina-election-a-referendum-on-sanford.html, accessed Sept. 14, 2014.

269 *Congressman texts explicit photos and messages:* Raymond Hernandez, "Weiner Resigns in Chaotic Final Scene," *New York Times,* June 16, 2011, at http://www.nytimes.com/2011/06/17/nyregion/anthony-d-weiner-tells-friends-he-will-resign.html?pagewanted=all; and CNN Staff, "Congress, Lewd Photos and NYC's Mayoral Race: An Anthony Weiner Timeline," *CNN Politics,* July 24, 2013, at http://www.cnn.com/2013/07/24/politics/weiner-timeline/, both accessed Sept. 14, 2014.

269 *Probable impact of Clinton scandal and impeachment:* Miller Center, University of Virginia, "Bill Clinton: Impact and Legacy," accessed Sept. 14, 2014, at http://millercenter.org/president/clinton/essays/biography/9.

270 *Catholic Church sexual abuse:* See "Roman Catholic Church Sex Abuse Cases," *New York Times,* Times Topics, April 18, 2014; for links to many articles, see http://topics.nytimes.com/top/reference/timestopics/organizations/r/roman_catholic_church_sex_abuse_cases/, accessed Sept. 14, 2014.

270 *Orthodox Jewish child molesters:* Sharon Ottoman and Ray Rivera, "Ultra-Orthodox Shun Their Own for Reporting Child Sexual Abuse," *New York Times,* May 9, 2012, at http://www.nytimes.com/2012/05/10/nyregion/ultra-orthodox-jews-shun-their-own-for-reporting-child-sexual-abuse.html?pagewanted=all, accessed Sept. 14, 2014.

270 *Major university football coach:* CNN Library, "Penn State Scandal Fast Facts," *CNN U.S.*, April 6, 2014, at http://www.cnn.com/2013/10/28/us/penn-state-scandal-fast-facts/, accessed Sept. 14, 2014. Sandusky's appeal of his verdict was denied by the Supreme Court of Pennsylvania on April 2, 2014.

270 *Horace Mann School cover-up:* Amos Kamil, "Prep-School Predators: The Horace Mann School's Secret History of Sexual Abuse," *New York Times,* June 6, 2012, http://www.nytimes.com/2012/06/10/magazine/the-horace-mann-schools-secret-history-of-sexual-abuse.html?pagewanted=all, accessed Sept. 14, 2014.

270 *Illinois governor went to jail:* Monica Davey, "Blagojevich Sentenced to 14 Years in Prison," *New York Times,* December 7, 2011, accessed Sept. 14, 2014, at http://www.nytimes.com/2011/12/08/us/blagojevich-expresses-remorse-in-courtroom-speech.html.

270 *Louisiana Congressman stored $90,000 in freezer:* Dave Cook, "Former Rep. William Jefferson Sentenced to 13 Years in Prison," *Christian Science Monitor,* November 13, 2009, accessed Sept. 14, 2014, at http://www.csmonitor.com/USA/Politics/2009/1113/former-rep-william-jefferson-sentenced-to-13-years-in-prison.

270 *New York Congressman censured for corruption:* David Kocieniewski, "Rangel Censured Over Violations of Ethics Rules," *New York Times,* December 2, 2010, at http://www.nytimes.com/2010/12/03/nyregion/03rangel.html?pagewanted=all, accessed Sept. 14, 2014.

270 *Martha Stewart went to jail:* Constance L. Hays, "Martha Stewart's Sentence: The Overview; 5 Months in Jail, and Stewart Vows, 'I'll Be Back,'" *New York Times,* July 17, 2004, accessed Sept. 14, 2004, at http://www.nytimes.com/2004/07/17/business/martha-stewart-s-sentence-overview-5-months-jail-stewart-vows-ll-be-back.html.

271 *Mladic on trial for mass murder:* "Ratko Mladic Trial Blocks Move to Drop Genocide Charges," *BBC News Europe,* April 15, 2014, accessed Sept. 14, 2014, at http://www.bbc.com/news/world-europe-27032446; and Allan Little, "Defiant Mladic Before War Crimes Court," *BBC News Europe,* May 16, 2012, accessed Sept. 14, 2014, at http://www.bbc.com/news/world-europe-18098463.

271 *Former Liberia president gets fifty years:* Owen Bowcott, "Charles Taylor's 50-Year Sentence Upheld at War Crimes Tribunal: Former Liberian President Is Now Likely to Be Sent to UK to Spend Rest of His Life in Jail," *Guardian,* September 26, 2013, at http://www.the

guardian.com/world/2013/sep/26/charles-taylor-liberian-president-sentence-upheld, accessed Sept. 14, 2014.

271 *Genocide in Darfur:* For links to many articles, see Times Topics, "Omar Hassan al-Bashir," *New York Times,* updated and accessed Sept. 14, 2014, at http://topics.nytimes.com/top/reference/timestopics/people/b/omar_hassan_al_bashir/index.html.

271 *Pol Pot:* Seth Mydans, "Death Of Pol Pot: Pol Pot, Brutal Dictator Who Forced Cambodians to Killing Fields, Dies at 73," *New York Times,* April 17, 1998, at http://www.nytimes.com/1998/04/17/world/death-pol-pot-pol-pot-brutal-dictator-who-forced-cambodians-killing-fields-dies.html, accessed Sept. 14, 2014.

271 *Slobodan Milosevic:* Molly Moore and Daniel Williams, "Milosevic Found Dead in Prison," *Washington Post,* March 12, 2006, accessed Sept. 14, 2014, at http://www.washingtonpost.com/wp-dyn/content/article/2006/03/11/AR2006031100525.html.

271 *Rwanda genocide:* "The ICTR Remembers: 20th Anniversary of the Rwandan Genocide," website of the International Criminal Tribunal for Rwanda, April 10, 2014, at http://unmict.org/ictr-remembers/, accessed Sept. 14, 2014.

271 *Central African Republic on the verge of genocide:* Josie Ensor, "Ban Ki-moon Warns of Rwandan Genocide Repeat in Central African Republic," *Telegraph,* April 6, 2014, accessed April 18, 2014, at http://www.telegraph.co.uk/news/worldnews/africaandindianocean/centralafricanrepublic/10747572/Ban-Ki-moon-warns-of-Rwandan-genocide-repeat-in-Central-African-Republic.html.

271 *"rape as a means of perpetrating genocide":* "The ICTR Remembers," http://unmict.org/ictr-remembers/, fully cited above. The self-description is quoted from the home page.

271 *Rape as genocide in Sudan:* David Scheffer, "Sudan and the ICC: Rape as Genocide," *New York Times,* November 3, 2008, accessed Sept. 14, 2014, at http://www.nytimes.com/2008/12/03/opinion/03iht-edscheffer.1.18365231.html.

273 *Women drivers:* Jason Kandel, "Insurance Study: Women Are Better Drivers Than Men," *NBC News,* January 12, 2012, accessed Sept. 14, 2014, at http://business.nbcnews.com/_news/2012/01/12/10142027-insurance-study-women-are-better-drivers-than-men; see also Anemona Hartocollis, "For Women Who Drive, the Stereotypes Die Hard," *New York Times,* August 17, 2010, at

http://www.nytimes.com/2010/08/18/nyregion/18drivers.html, accessed Sept. 14, 2014.

274 *"Women in Sports": Time,* July 26, 1978. To see the cover, go to http://content.time.com/time/covers/0,16641,19780626,00.html, accessed Sept. 14, 2014. A link to the article "Comes the Revolution: Joining the Game at Last, Women Are Transforming American Athletics" is provided there. For the Mia Hamm photo, visit http://www.values.com/inspirational-sayings-billboards/56-passion; another example is here: http://academic.reed.edu/anthro/faculty/mia/Images/Gallery/Pics/soccerwoman.jpg, both accessed Sept. 14, 2014.

274 *Women in sports statistics and history:* R. Vivian Acosta and Linda Jean Carpenter, "Women in Intercollegiate Sport: A Longitudinal, National Study, Thirty Seven Year Update, 1977–2014" (unpublished manuscript), available for downloading at www.acostacarpenter.org, accessed Sept. 14, 2014. The estimates quoted here are based on data on p. 1.

274 *Women at West Point:* Larry Abramson, "West Point Women: A Natural Pattern or a Camouflage Ceiling?" *Morning Edition,* National Public Radio (NPR) website, October 22, 2013, at http://www.npr.org/2013/10/22/239260015/west-point-women-a-natural-pattern-or-a-camouflage-ceiling, accessed Sept. 14, 2014.

274 *Women in combat roles:* See, for example, the *Christian Science Monitor*'s photo gallery at http://www.csmonitor.com/Photo-Galleries/In-Pictures/Women-in-Combat#514100, accessed Sept. 14, 2014; see also CNN Staff, "Pentagon Says Women in All Combat Units by 2016," *CNN Politics,* June 18, 2013, accessed Sept. 14, 2014, at http://www.cnn.com/2013/06/18/politics/women-combat/.

275 *The majority graduating are women:* Christopher B. Swanson, "U.S. Graduation Rate Continues Decline," *Education Week,* June 2, 2010, accessed Sept. 14, 2014, at http://www.edweek.org/ew/articles/2010/06/10/34swanson.h29.html; "College Enrollment and Work Activity of 2012 High School Graduates," Bureau of Labor Statistics, U.S. Department of Labor, April 17, 2013, accessed Sept. 14, 2014, at http://www.bls.gov/news.release/hsgec.nr0.htm; "Digest of Education Statistics," table 258, National Center for Education Statistics, accessed Sept. 14, 2014, at http://nces.ed.gov/programs/digest/d07/tables/dt07_258.asp.

275 *Achievement especially high in first-born girls:* Katy Waldman, "Firstborn Girls Are Statistically More Likely to Run the World,"

*Slate*, April 28, 2014, accessed Sept. 23, 2014, at http://www.slate.com/blogs/xx_factor/2014/04/28/firstborn_girls_are_more_accomplished_and_ambitious_than_firstborn_boys.html; and Feifei Bu, "Sibling Configurations, Educational Aspiration and Attainment," *ISER Working Paper Series* no. 2014–11, Institute for Social and Economic Research, University of Essex, February 2014.

275 *women regained 95 percent of lost jobs:* "Women Sharing in Slow Job Growth, but Full Recovery Is a Long Way Off," National Women's Law Center website, August 2, 2013, at http://www.nwlc.org/resource/women-sharing-slow-job-growth-full-recovery-long-way#_edn1, accessed Sept. 14, 2014.

276 *the number of women going into business:* WomenAble, "The 2014 State of Women-Owned Businesses Report," prepared by Womenable, commissioned by American Express OPEN, Executive Summary issued April 2, 2014, at http://www.womenable.com/59/the-state-of-women-owned-businesses-in-the-us:-2014, accessed Sept. 14, 2014; see also "Women-Owned Businesses," National Women's Business Council, http://www.nwbc.gov/facts/women-owned-businesses, accessed Sept. 14, 2014.

276 *"Entrepreneurship is the new women's movement":* Natalie MacNeil, "Entrepreneurship Is the New Women's Movement," Forbes.com, accessed Sept. 14, 2014, at http://www.forbes.com/sites/work-in-progress/2012/06/08/entrepreneurship-is-the-new-womens-movement/.

276 *"I think it was just a matter of time":* Sheena MacKenzie and Isha Sesay, "Helene Gayle: Meet the CEO Feeding the World," CNN Leading Women, September 17, 2013, at http://edition.cnn.com/2013/09/17/business/helene-gayle-meet-the-ceo/, accessed Sept. 14, 2014.

276 *Women lead around 19 percent of charities:* Heather Joslyn, "A Man's World," *Chronicle of Philanthropy,* September 17, 2009, accessed Sept. 14, 2014, at https://philanthropy.com/article/A-Mans-World/57099/.

276 *Women at Harvard Business School:* Jodi Kantor, "Harvard Business School Case Study: Gender Equity," *New York Times,* September 7, 2013, accessed Sept. 14, 2014, at http://www.nytimes.com/2013/09/08/education/harvard-case-study-gender-equity.html?pagewanted=all&_r=0.

277 *"It's, like, perfect . . . my trophy husband":* Video by Brent McDonald captioned "Painful Experience," on the web page cited in the entry above, accessed Sept. 14, 2014.

277 *Kantor on WNYC:* "Gender and Class at Harvard Business School," *Brian Lehrer Show,* September 13, 2013, WNYC website, accessed Sept. 14, 2014, at http://www.wnyc.org/story/317858-gender-and-class-harvard-business-school/.

278 *Women's political attitudes:* Torben Iversen and Frances Rosenbluth, "The Political Economy of Gender: Explaining Cross-National Variation in the Gender Division of Labor and the Gender Voting Gap," *American Journal of Political Science* 50, no. 1 (2006): 1–19; A. H. Eagly, A. B. Diekman, M. C. Johannesen-Schmidt, and A. M. Koenig, "Gender Gaps in Sociopolitical Attitudes: A Social Psychological Analysis," *Journal of Personality and Social Psychology* 87, no. 6 (2004): 796–816.

278 *Women and men helpful in different ways:* Alice H. Eagly, "The His and Hers of Prosocial Behavior: An Examination of the Social Psychology of Gender," *American Psychologist* 64, no. 8 (2009): 644–58.

278 *Women "support the provision of social services":* Eagly et al., "Gender Gaps," 798.

279 *"Women also advocate more restriction of many behaviors":* Ibid.

279 *The gender voting gap in ten countries:* Iversen and Rosenbluth, "Political Economy of Gender."

280 *"Given the overall trend":* Ibid., 18.

280 *"In the latter part of the 20th century":* Paula Wyndow, Jianghong Li, and Eugen Mattes, "Female Empowerment as a Core Driver of Democratic Development: A Dynamic Panel Model from 1980 to 2005," *World Development* 52 (2013): 34.

281 *"Rather than being a natural consequence":* Ibid., 47.

281 *"In a Senate still dominated by men":* Jonathan Weisman and Jennifer Steinhauer, "Senate Women Lead in Effort to Find Accord," *New York Times,* October 14, 2013. The next few quotes, through the Susan Collins quote, are from the same article.

282 *"My own experience in Congress":* Marianne Schnall, "Conversation with Senator Kirsten Gillibrand," Feminist.com interview with New York senator Kirsten Gillibrand, September 20, 2011, accessed Sept. 14, 2014, at http://www.feminist.com/resources/artspeech/interviews/kirstengillibrand.html.

282 *Senators leading compromise have children:* This information was accessed on Sept. 14, 2014. For Kelly Ayotte: http://www.rollcall

.com/members/32609.html; Lisa Murkowski: http://www.rollcall.com/members/17262.html; Patty Murray: http://media.cq.com/members/522/rc=1; Amy Klobuchar: http://www.rollcall.com/members/25668.html; Jeanne Shaheen: http://www.rollcall.com/members/1444.html.

283 *Nineteen women Senators interviewed by Sawyer:* The quotations from Mikulski, Feinstein, Gillibrand, Ayotte, and McCaskill are from excerpts from the interview, accessed Sept. 14, 2014, at http://abcnews.go.com/Politics/meet-class-senate-swears-historic-20-female-senators/story?id=18113363. The video is remarkable.

283 *"Female mayors were far more willing to change":* Lynne A. Weikart, Greg Chen, Daniel W. Williams, and Haris Hromic, "The Democratic Sex: Gender Differences and the Exercise of Power," *Journal of Women, Politics & Policy* 28, no. 1 (2006): 119.

283 *"The joy of having power":* Ibid., 120.

284 *"Frequently mothers will introduce me to their daughters":* Ibid., 133.

284 *"She has been a rank-and-file engineer":* Bill Vlasic, "New G.M. Chief Is Company Woman, Born to It," *New York Times,* December 10, 2013, print edition, p. 1; also at http://www.nytimes.com/2013/12/11/business/gm-names-first-female-chief-executive.html, accessed Sept. 14, 2014.

284 *"Mary was picked for her talent":* Ibid.

285 *"Things have changed dramatically":* "Mary Barra, G.M.'s New Chief, Speaking Her Mind," *New York Times,* December 10, 2013, accessed Sept. 14, 2014, at http://www.nytimes.com/2013/12/11/business/mary-barra-gms-new-chief-speaking-her-mind.html?ref=business.

285 *Barra's response to ignition-system failure:* Bill Vlasic, "Recall at G.M. Is Early Trial for New Chief," *New York Times,* March 7, 2014, accessed Sept. 14, 2014, at http://www.nytimes.com/2014/03/08/business/recall-at-gm-is-early-trial-for-new-chief.html.

285 *"a sweeping internal investigation":* Bill Vlasic, "G.M. Inquiry Cites Years of Neglect over Fatal Defect," *New York Times,* June 5, 2014, A1, at http://www.nytimes.com/2014/06/06/business/gm-ignition-switch-internal-recall-investigation-report.html?module=Search&mabReward=relbias%3Aw&_r=0, accessed Nov. 4, 2014.

286 *The first car to drive itself:* Nathan Bomey, "Barra commits GM to car-to-car link," *Detroit Free Press*, accessed Sept. 14, 2014 at *USA Today*

website, http://www.usatoday.com/story/money/cars/2014/09/07/gm-connected-car-barra/15245409/.

286 *"a sampling of male and female leaders":* Jack Zenger and Joe Folkman, "A Study in Leadership: Women Do It Better Than Men" (Orem, UT: Zenger/Folkman, 2012), 1, downloaded April 19, 2014, from http://www.zfco.com/media/articles/ZFCo.WP.WomenBetterThanMen.033012.pdf.

286 *"average rating from an aggregate":* Ibid., 2.

287 *"It is a well-known fact":* Quoted in "Zenger Folkman: Women Score Higher in Majority of Leadership Competencies," Business Wire: A Berkshire Hathaway Company, March 15, 2012, accessed Sept. 14, 2014, at http://www.businesswire.com/news/home/20120315006216/en/Zenger-Folkman-Women-Score-Higher-Majority-Leadership#.U1Ksxtw0dCg.

287 *"Sizable differences . . . in employment policies":* David A. Matsa and Amalia R. Miller, "A Female Style in Corporate Leadership? Evidence from Quotas," *American Economic Journal: Applied Economics* 5, no. 3 (2013): 138.

288 *"Ultimately, time will tell":* Ibid., 153.

288 *European parliament corporate gender quota:* Charlotte McDonald-Gibson, "'The First Cracks in the Glass Ceiling': EU Votes to Impose Legal Quotas for Women in the Boardroom," *Independent,* November 20, 2013, accessed Sept. 14, 2014, at http://www.independent.co.uk/news/world/europe/the-first-cracks-in-the-glass-ceiling-eu-votes-to-impose-legal-quotas-for-women-in-the-boardroom-8952718.html.

288 *the 2014 publication of an American study:* Matsa and Miller, "Reductions at Women-Owned Businesses in the United States," *Industrial and Labor Relations Review* 67, no. 2 (2014): 422–52.

288 *"workforce reductions are more than twice as frequent":* Ibid., 432.

288 *"It's just a terrible thought":* Leibinger-Kammüller quoted by Matsa and Miller, 422.

288 *"Women's empathy enables them to look":* Spencer quoted by Matsa and Miller, 422.

289 *"Male directors care more about achievement":* Renée B. Adams and Patricia Funk, "Beyond the Glass Ceiling: Does Gender Matter?" *Management Science* 58, no. 2 (2012): 220.

289 *"Female directors are less security oriented":* Ibid.

290 *"Sex segregation in corporate criminality is pervasive":* D. J. Steffensmeier, J. Schwartz, and M. Roche, "Gender and Twenty-First-Century Corporate Crime: Female Involvement and the Gender Gap in Enron-Era Corporate Frauds," *American Sociological Review* 78, no. 3 (2013): 448.

290 *Parliamentary gender balance and national corruption:* David Dollar, Raymond Fisman, and Roberta Gatti, "Are Women Really the 'Fairer' Sex? Corruption and Women in Government," *Journal of Economic Behavior & Organization* 46 (2001): 423–29.

290 *"a substantial literature"* and *"our results suggest":* Ibid., 427.

291 *Women in African parliaments:* Gretchen Bauer, "Gender Quotas and Women's Representation in African Parliaments," Democracy in Africa website, accessed Sept. 14, 2014, at http://democracyinafrica.org/gender-quotas-womens-representation-african-parliaments/.

## Epilogue: #YesAllWomen

295 *"war on women"* and *"Girls have never been attracted to me":* Quotations and video available in "#YesAllWomen: Rebecca Solnit on the Santa Barbara Massacre & Viral Response to Misogynist Violence," *Democracy Now!* website, May 27, 2014, accessed Sept. 14, 2014, at http://www.democracynow.org/2014/5/27/yesallwomen_rebecca_solnit_on_the_santa. This video was taken down by YouTube, but as of Sept. 14, 2014, it was available with a warning: https://www.youtube.com/watch?v=MQUW3Km01BM&bpctr=1402426134. His written self-justification and plan for systematic mass murder can be found (also with a warning) on the website of *New York Times,* accessed Sept. 14, 2014, at http://www.nytimes.com/interactive/2014/05/25/us/shooting-document.html.

296 *"We have an abundance of rape":* Rebecca Solnit, *Men Explain Things to Me* (Chicago: Haymarket Books, 2014), 21; also quoted in the interview cited above.

296 *How #YesAllWomen came about:* Jennifer Medina, "Campus Killings Set Off Anguished Conversation About the Treatment of Women," *New York Times,* May 26, 2014, http://www.nytimes.com/2014/05/27/us/campus-killings-set-off-anguished-conversation-about-the-treatment-of-women.html, accessed Sept. 14, 2014.

296 *"#YesAllWomen because 'I have a boyfriend'":* These three tweets are quoted by Sasha Weiss, "The Power of #YesAllWomen," *New Yorker* website, May 26, 2014, at http://www.newyorker.com/online/blogs/culture/2014/05/the-power-of-yesallwomen.html, accessed Sept. 14, 2014.

296 *"Because what men fear most":* These three tweets or retweets posted in early June were found on the Twitter website at https://twitter.com/hashtag/YesAllWomen. All three could also be found by searching twitter.com on Nov. 4, 2014.

297 *Nuclear transfer approved for United Kingdom:* Steve Connor, "UK Becomes First Country in World to Approve IVF Using Genes of Three Parents," *Independent,* June 28, 2013, at http://www.independent.co.uk/news/science/uk-becomes-first-country-in-world-to-approve-ivf-using-genes-of-three-parents-8677595.html, accessed Sept. 14, 2014.

298 *The Y chromosome has shrunk in size:* Doris Bachtrog, "Y-Chromosome Evolution: Emerging Insights into Processes of Y-Chromosome Degeneration," *Nature Reviews Genetics* 14, no. 2 (2013): 113–24.

298 *a 2014 study suggests it can't shrink much further:* M. A. Wilson Sayres, K. E. Lohmueller, R. Nielsen, "Natural Selection Reduced Diversity on Human Y Chromosomes," *PLoS Genetics* 10, no. 1 (2014): e1004064.

298 *Lionel Tiger's* The Decline of Males*:* Lionel Tiger, *The Decline of Males* (New York: Golden Books, 1999).

298 *France research shows sperm count decline:* M. Rolland, J. Le Moal, V. Wagner, D. Royere, and J. De Mouzon, "Decline in Semen Concentration and Morphology in a Sample of 26,609 Men Close to General Population between 1989 and 2005 in France," *Human Reproduction* 28, no. 2 (2013): 462–70.

298 *Finland decline:* N. Jorgensen, M. Vierula, R. Jacobsen, E. Pukkala, A. Perheentupa, H. E. Virtanen, N. E. Skakkebk, and J. Toppari, "Recent Adverse Trends in Semen Quality and Testis Cancer Incidence Among Finnish Men," *International Journal of Andrology* 34, no. 4 (2011): E37–E48.

298 *Denmark decrease in sperm quality:* Niels Jorgensen and eleven other authors, "Human Semen Quality in the New Millennium: A Prospective Cross-Sectional Population-Based Study of 4867 Men," *BMJ Open* 2, no. 4 (2012).

299 *India study:* Madhukar Shivajirao Dama and Singh Rajender,

"Secular Changes in the Semen Quality in India During the Past 33 Years," *Journal of Andrology* 33, no. 4 (2012): 740–44.

299 *About 40 percent of births are to unmarried women:* Brady E. Hamilton, Joyce A. Martin, and Stephanie J. Ventura, "Births: Preliminary Data for 2012," *National Vital Statistics Reports* 62, no. 3, September 6, 2013, accessed Sept. 14, 2014, at http://www.cdc.gov/nchs/data/nvsr/nvsr62/nvsr62_03.pdf.

299 *Children will spend some time in single-parent households:* Gunnar Andersson, "Children's Experience of Family Disruption and Family Formation: Evidence from 16 FFS Countries," *Demographic Research* 7 (2002): 343–64. The United States leads in these statistics, although Latvia is a close second.

299 *Married women who outearn husbands:* Catharine Rampell, "U.S. Women on the Rise as Family Breadwinner," *New York Times,* May 29, 2013, http://www.nytimes.com/2013/05/30/business/economy/women-as-family-breadwinner-on-the-rise-study-says.html?_r=1&, accessed Sept. 14, 2014.

300 *Children with two mothers or two fathers:* Charlotte J. Patterson, "Children of Lesbian and Gay Parents: Psychology, Law and Policy," *American Psychologist* 64, no. 8 (2009): 727–36. See also Konner, *Evolution of Childhood,* 329–34.

300 *Complex impact of divorce on children:* E. Mavis Hetherington, "Divorce and the Adjustment of Children," *Pediatrics in Review* 26, no. 5 (2005): 163–69; E. M. Hetherington, "Social Support and the Adjustment of Children in Divorced and Remarried Families," *Childhood: A Global Journal of Child Research* 10, no. 2 (2003): 217–36.

300 *Children's resilience and protective factors:* Michael Rutter, "Annual Research Review: Resilience—Clinical Implications," *Journal of Child Psychology and Psychiatry* 54, no. 4 (2013): 474–87; E. E. Werner, "Vulnerable but Invincible: High-Risk Children from Birth to Adulthood," supplement, *Acta Paediatrica* 422 (1997): 103–05.

304 *"The world's gone social":* Sandberg is quoted by Jenna Goudreau in "What Men and Women Are Doing on Facebook," which summarizes male-female differences in social media use: *Forbes,* April 26, 2010, accessed Sept. 14, 2014, at http://www.forbes.com/2010/04/26/popular-social-networking-sites-forbes-woman-time-facebook-twitter.html.

304 *"A woman is the child of":* Laurie Goering, "In Lesotho, Women

Hope for Control of Their Lives," *Chicago Tribune,* October 17, 2004, accessed Sept. 14, 2014, at http://articles.chicagotribune.com/2004-10-17/news/0410170377_1_lesotho-husband-s-permission-south-africa.

305 *"Like a glacier":* Helen Fisher, *The First Sex: The Natural Talents of Women and How They Are Changing the World* (New York: Ballantine, 2000), 288.

# For Further Reading

Important predecessors of this book or some critical aspects of it are Margaret Mead's *Male and Female*, Ashley Montagu's *The Natural Superiority of Women*, Simone de Beauvoir's *The Second Sex*, Sarah Blaffer Hrdy's *The Woman That Never Evolved*, Laura Betzig's *Despotism and Differential Reproduction*, Camille Paglia's *Sexual Personae*, Helen Fisher's *The First Sex*, Bobbi Low's *Why Sex Matters*, Steven Pinker's *The Better Angels of Our Nature*, and Hanna Rosin's *The End of Men*. This should not be taken to mean that these authors would agree with all my claims (I don't agree with all of theirs), just that I have learned a lot from them that helped me write this book. On a sadder note and one on which there is no disagreement on my part, Nicholas Kristof and Sheryl WuDunn's *Half the Sky* is a broad and moving overview of what is being done to help oppressed women throughout the world.

For an entertaining and reliable account of the evolution of sex, see Jared Diamond's *Why Is Sex Fun?* And for the full range of animal mating systems and sexual proclivities, you can't do better than the delightful, hilarious, and completely authoritative *Dr. Tatiana's Sex Advice to All Creation*, by Olivia Judson. Dr. T lets you know what you must do to succeed in the mating game if you are a fairy wren, a golden potto, a cockroach, or one of any number of other creatures facing the eternal puzzles of love and life.

An extended treatment of sexual orientation and identity in all their manifestations was not in the scope of this book. Joan Roughgarden's *Evolution's*

*Rainbow* (although it contains an immoderate attack on the theory of sexual selection) is a fascinating and celebratory account of the variety of animal and human sex and gender arrangements, by a distinguished biologist who is herself an openly transgender woman with a moving personal story. It should be read in connection with the reviews in *Nature* by Sarah Blaffer Hrdy and in *Science* by Alison Jolly. Gilbert Herdt's *Third Sex, Third Gender* remains the definitive source on cross-cultural variation in roles for people who do not fit easily into the categories usually labeled "male" and "female." One of the most illuminating—in fact, riveting—first-person accounts of gender is Norah Vincent's *Self-Made Man,* in which she describes with great insight and compassion what she saw and learned in a year of successfully impersonating a man.

Although I see *Women After All* as a brief in favor of women, some will see it as a brief against men. To some extent that is inevitable, for two reasons. First, along with their many accomplishments, men have done a great deal of damage. Second, if women are superior even in some ways, then by definition men must be inferior in those ways. But I have said that there is good reason to worry about boys and men. It can be simultaneously true that (1) men have held and wielded most of the power in history and that (2) most men are not now and never were powerful. Other books have given attention to these facts, as well as to the particular vulnerabilities boys and men face as women's power grows. Among these I recommend Warren Farrell's *The Myth of Male Power,* Christina Hoff Sommers's *The War Against Boys,* Lionel Tiger's *The Decline of Males,* and Roy Baumeister's *Is There Anything Good About Men?* Camille Paglia can always be counted on for a spirited and sophisticated defense of men, through easily available essays, speeches, and interviews (for example, in *Time,* December 30, 2013). I think of these views as a needed antidote to an idea I have not advanced and do not share: that men are all-powerful and mostly without value.

# Index

# About the Author

Melvin Konner is the Dobbs Professor of Anthropology and Behavioral Biology at Emory University. After attending Brooklyn College, he received his Ph.D. and, later, his M.D. from Harvard, where he taught before moving to Emory. For two years he lived with and studied the !Kung San hunter-gatherers of Botswana. Aside from his ten books (including *The Tangled Wing: Biological Constraints on the Human Spirit, Becoming a Doctor,* and *The Evolution of Childhood*), he has written for the *New York Times, Newsweek, Psychology Today,* and the *New York Review of Books,* as well as *Nature, Science,* the *New England Journal of Medicine, Child Development,* and other journals. He has held fellowships from the Guggenheim and Fulbright foundations and the Center for Advanced Study in the Behavioral Sciences. He has testified at U.S. Senate hearings on health reform and end-of-life issues; has given distinguished lectures on medical humanities at the Yale, Mayo Clinic, and Vanderbilt medical schools; and has lectured on evolutionary medicine, the evolution of childhood, and other topics at universities around the world. A fellow of the American Association for the Advancement of Science, he has won the American Anthropological Association's Anthropology in Media Award and is listed in *Who's Who in America* as well as *Who's Who in Hell.* He has four grown children—three daughters and a son—and was for a decade a single father. His website is www.melvinkonner.com.